# HUMAN FACTORS IN AIRCRAFT MAINTENANCE

# HUMAN FACTORS IN AIRCRAFT MAINTENANCE

Demetris Yiannakides and
Charalampos Sergiou

**CRC Press**
Taylor & Francis Group
Boca Raton London New York

CRC Press is an imprint of the
Taylor & Francis Group, an **informa** business

CRC Press
Taylor & Francis Group
6000 Broken Sound Parkway NW, Suite 300
Boca Raton, FL 33487-2742

© 2020 by Taylor & Francis Group, LLC

CRC Press is an imprint of Taylor & Francis Group, an Informa business

**Library of Congress Cataloging-in-Publication Data**

Names: Yiannakides, Demetris, author. | Sergiou, Charalampos, author.
Title: Human factors in aircraft maintenance / authored by Demetris Yiannakides
    and Charalampos Sergiou.
Description: Boca Raton : CRC Press, 2019. | Includes bibliographical references and
    index.
Identifiers: LCCN 2019025505 (print) | LCCN 2019025506 (ebook) |
    ISBN 9780367230111 (hardback : acid-free paper) | ISBN 9780429280887 (ebook)
Subjects: LCSH: Aeronautics–Human factors. | Airplanes–Maintenance and
    repairl. | Aircraft accidents–Human factors–Case studies. | Aircraft accidents–
    Investigation–Case studies.
Classification: LCC TL671.9 .Y53 2019 (print) | LCC TL671.9 (ebook) | DDC
    629.134/6–dc23
LC record available at https://lccn.loc.gov/2019025505
LC ebook record available at https://lccn.loc.gov/2019025506

**Visit the Taylor & Francis Web site at**
www.taylorandfrancis.com

**and the CRC Press Web site at**
www.crcpress.com

# Dedication

*To our parents and lovely families*
*To all those who envision high levels of aviation safety*

# Contents

# List of Figures

# List of Tables

# Preface

This book, *Human Factors in Aircraft Maintenance*, has two goals. First of all, to be a competent and efficient academic text in the hands of trained engineers and their instructors and, second, to constitute a work permit for licensed aircraft engineers and their management, which will train them on the causes and impact of failing human factors on personal health and aircraft safety. Concurrently, this book incorporates a proposal for the revision of the EU Commission Regulation 1321/2014, and in particular Annex III (Part-66) and Annex IV (Part-147), which is envisaged to be adopted by other aviation authorities and agencies around the world. The overall purpose of the proposed revision is to update the existing initial training system of aircraft maintenance engineers in a way that will deliver skilled and well-educated engineers to the industry in the challenging years ahead, who will consciously adopt a high level of safety culture and remain unaffected by stressful conditions imposed in the workplace.

To achieve this above, the book incorporates the accumulated experience and the scientific and research point of view that is evident throughout journals, books and conference proceedings. It also focuses on the human factor as the latent cause for a number of recorded aircraft accidents and incidents, without any intention to identify any personal or organizational fault or liability, which is neither the function nor the intention of the authors. Rather, the aim is to point out that the principles of human factors theory and the safety theoretical models apply to real events and consequently are crucial in preventing future accidents and incidents. Finally, this book highlights the lessons learned and points out the organizational and individual responsibilities. In addition, it raises the accountability issues associated with the profession by concisely distinguishing human error types.

# Acknowledgments

The idea of preparing this book was born during our visit as instructors to the Ikaros Training Center, an EASA part-147 organization, where we realized the shortcomings of the existing education/training system, as well as the shortage of a single book which could concentrate the scattered academic, technical and empirical knowledge of Human Factors in Aircraft Maintenance.

Our commitment to the successful achievement of our goal was driven from the start by our love for aviation and our strong aspiration to provide high-quality knowledge to aircraft maintenance engineers. Our goal, which remains unchanged, is the improvement of safety culture through continuous training, for the influence of the human factor in the daily work of engineers and the impact it has on their personal safety and the safety of aircraft.

In addition to our families, this endeavor was firmly supported by Christos Christodoulou, head of the Air Force Flight Safety Directorate, whom we thank for countless hours of discussion and exchange of views, and Andreas Gennaris who undertook the task of upgrading the written discourse and maintaining consistency throughout this book.

Finally, it would be an omission not to thank our publishers for believing in us and in our common goal.

# About the Authors

**Demetris Yiannakides** graduated from the Hellenic Air Force Academy as an aircraft pilot, receiving his Bachelor in Aviation Science, and then continued post-graduate studies with a Masters in Aeronautical Science at Embry Riddle Aeronautical University. Further to his everyday duties as an instructor pilot, he is also a visiting instructor in an EASA Part-147 Training center in the fields of Human Factors and Aerodynamics. In parallel, he often participates in air navigation services audits as a qualified auditor. He and his wife live in Nicosia, Cyprus, with one daughter. He enjoys watching football and drinking coffee with long-lasting friends. Deeply affected by the loss of colleagues in the line of duty, he has devoted himself to raising awareness concerning the dramatic impact of failing human factors on aviation safety.

**Dr. Charalampos Sergiou** graduated from the Hellenic Air Force Academy as a Telecommunications and Electronics Engineer. He holds an MSc in Advance Information Technology and a PhD in Wireless Ad-Hoc Networks, from the Department of Computer Science of the University of Cyprus. He is an Aviation Engineer in the Cyprus Air Force and EASA Part-66 Licensed Aviation Maintenance Engineer (LAME). He holds a B1 (Full) and B2 License and he is type rated (B2) in AW139 Helicopters and Airbus 318/319/320/321 (CFM-56, V2500) aircraft types. He is also an EASA Part-147 Instructor and Head of Examinations (Form 4 Holder) in an EASA Part-147 Training Center in Cyprus. He has published several articles in International Conferences and Journals (IEEE, Elsevier, ACM etc.) where he also serves as a reviewer. He and his wife live in Nicosia, Cyprus, with their two daughters.

# 1 Introduction

Air transportation emerged in the middle of the last century as the latest mode of public transportation (Dempsey et al. 1997) and since then has been experiencing continuous growth. Its driving powers have been the remarkable evolution of technology, the strong resilience of the industry promising a sustainable future (IATA 2018b; Macário et al., 2009) against any global economic downturns (Goyal and Negi 2014; European Parliament 2009), the unprecedented safety performance (Boeing 2017a; Airbus 2018) and the public's constant loyalty to air transportation (IATA 2018b). Now more than ever, the industry is characterized by fierce competition and a vast demand for time aircraft utilization, which inevitably generates conditions that drive human capabilities and performance to the edge. Aircraft maintenance is an integral part of the aviation system that experiences the direct implications of the enlargement of the industry. In effect, aircraft engineers frequently work under harsh conditions to the utmost of their physiology and mental capacity by undertaking tasks in complex systems under accumulated workload and pressure, knowing that a single error may deny the delivery of an airworthy aircraft.

At the same time, it is of no surprise that the International Civil Aviation Organization (ICAO), national authorities and other competent aviation agencies, having examined the lessons learned from recent research studies and accident/incident investigation reports, continue to identify human factors as the most frequent, primary causal factor and/or the main contributor to aviation accidents and incidents (CAA UK 2015; CASA Australia 2013). Statistics show that about 70–80 percent of all aviation accidents across the globe are caused by human error, the pilot error problem having a higher profile. Some 20 percent of these errors are estimated to occur due to maintenance activities that may contain either failed components and/or components involved in procedures that do not comply with regulations, policies and processes (Boeing 2013). Consequently, authorities, in an effort to raise the awareness of the industry regarding the implications of human failure in aviation safety and to avoid complacency have issued regulations imposing a compulsory theoretical module in human factors during the initial training of maintenance engineers, training that regularly repeats during the course of their career in the aviation industry (EU Commission Regulation 2014; CASA Australia 1998–2018; Civil Aviation Administration of China 2005). Identically, they have issued recommendations and non-mandatory advisory circulars for human factors training (FAA 2017) or have introduced standard apprenticeship (Ministry of Civil Aviation of India 2016) within the maintenance organizations, amid open discussions for integrating human factors training in the existing regulation frameworks, to ensure the delivery of competent, skilled and safe personnel (Johnson 2018).

Fundamentally, human factors science incorporates the fields of psychology, physiology, sociology, engineering, industrial design, statistics, operations research and anthropometry, and provides an understanding of the impact of human limitations over human performance. To this end, human factors examine the relationship between the people and the components of the systems in terms of interoperability and consistency. Ultimately, the acquired knowledge, once complemented with the ability to recognize and master the dangers that stem from human behavior, may act as the enabler towards the mitigation of human failure and the design of error tolerant and more resilient systems. Notably, the Aviation European Human Factors Advisory Group (EHFAG) in its *European Strategy for Human Factors in Aviation* suggests that training in human factors considers error as a symptom of systemic and organizational issues, with multiple factors involved that affect human performance (2012).

Human factors knowledge not only enhances human performance and facilitates error mitigation but contributes to the goal of creating and reinforcing a *safety culture* within organizations where employees practice safe habits (FAA 2014). The UK's Health and Safety Executive (HSE) defines safety culture as the product of individual and group values, attitudes, perceptions, competencies and patterns of behavior that determine the commitment to safety (1993). To this end, organizations promote safety culture by considering their inherent safety values, leadership strategies, employee attitudes and behaviors (Salas and Maurino 2010; Patterson 2002). Existing efforts converge in integrating safety culture into corporate culture through the application of safety models and safety management systems (ICAO 2018). Indisputably, safety culture is captured very high within the taxonomy of safety drivers and remains a significant feature that should characterize every organization functioning in the aviation industry.

Concurrently, the forecasts of major aircraft manufacturers (Boeing 2017b; Airbus 2017), as a well as those of the International Civil Aviation Organization (ICAO 2017), reveal that industry should not be left in complacency, since they expect that the world commercial fleet will more than double over the next 20 years. Boeing estimates that demand in the commercial market for jet airplanes will nearly double to 47,000 airplanes, at an average annual growth rate of 3.3 percent. The same is true for Airbus, who forecasts a demand for 34,900 aircraft. Amazingly, more than 600,000 aircraft engineers will need to be trained at the highest levels (Boeing 2017a) and employed in the industry until 2036.

It is indeed worrying that the estimated enlargement of the industry and the consequent tremendous need for aircraft engineers comes with significant industry complaints, that even though human factors are being gradually integrated into safety cultures and safety management systems incidents manifesting human errors still occur. An aggravating factor is also the findings of studies that expect a labor shortage in the maintenance technician field in the coming years, with the most alarming finding to be the problem of skills shortage of the new engineers employed in aircraft maintenance (Lewis 2012; Karlsson 2007; US Government Accountability Office 2014).

It is the view of the authors that the training organizations in their effort to satisfy the demand for aircraft engineers, and due to the fact that the current training system is not as standardized as it should be, allowing for variations in the quality of the delivered training will inevitably have a significant negative impact on the skills of the new engineers. The shortfalls and the potential implications in aircraft maintenance converge on the view that the aviation community must strive for lasting solutions to face the challenges. It is therefore suggested that collective efforts should focus on the entire current initial training of engineers and the relevant curricula and training methodologies.

A significant pitfall under the existing training is that at the initial training, there is not a structured methodology in teaching human factors within the working environment, a methodology capable of enhancing situational awareness and practicing advanced skill behavior techniques when dealing with multitasks and complex systems. Engineers engage in developing safety culture, initially within a mature working environment (part 145/M-station or base repair organisations), which needs to deliver airworthy aircraft under significant stressors such as time pressure and heavy workload and where there is no opportunity for exercising, planning or practicing different scenarios and techniques. The second and equally important parameter is the fact that the key front-line personnel usually perceive safety culture as a gospel of regulations required to be implemented as a checklist. In addition, the fact should not be missed that top-level managers responsible for promoting a safety culture within organizations by improving behaviors and attitudes do not necessarily possess a sound understanding of safety culture themselves. Another reason is the existing confusion that surrounds the concepts of safety culture and health and safety rules. It is a requirement that this relationship should be entirely distinct.

In that sense, we propose that safety attitudes, improved skills, valued performance and the foundations of safety culture can be better adopted and built during the initial training of the aircraft engineers within the training institutions and organizations. At initial training where "fermentation" takes place, instructors should be obliged by the training system to simulate the harsh conditions of the "true" working environment with sufficient realism and vaccinate young engineers with the values of safety culture. At this very important learning stage, as Bainbridge suggests, "all the expertise of psychology learning [...] could be included [...] any type of behavior and any mode of cognitive processing can be skilled" (1999). This long-lasting positive attitude that will be consciously acquired through knowledge and practice and further enriched by increasing experience could be used as a predictive tool of human behavior over a task and positively influence the already established maintenance teams (Choi and Levine 2004) without being dependent on other endogenous and exogenous factors or rely on the situation.

The proposed training approach is holistic and takes into account the theory of human factors as well as the specificities of the character, the personality and the national culture of each young aircraft engineer. As Depaolo and Mclaren suggest, by improving an individual's attitude we also increase human learning ability (2006), which is a plus to the common understanding of human failure and risk management strategies, which in today's approach remains mainly theoretical. In effect, the results of the new approach should satisfy the needs of the industry and also fortify the efforts of the aviation authorities and maintenance organizations. The goal is to eliminate unsafe actions and risky behaviors that inevitably harm personnel's health and jeopardize flight safety. Such unsafe behaviors cannot be the rule but the absolute exception.

## 1.1 THE NEED TO TAKE HUMAN FACTORS INTO ACCOUNT: A HISTORICAL PERSPECTIVE

From the origins of air transport to the present day, the perception that prevailed, as well as the methodology used to ensure safety, has undergone considerable variations due to the evolution of air transport itself. Its determinant factor has been the aftermath of accidents and the introduction of innovative investigation techniques, which gradually revealed the overall sequence of causal factors and demonstrated the need for effective and revolutionary mitigation measures. As it is shown in Figure 1.1, the evolution in aviation safety thinking can be divided into four consecutive eras (ICAO 2018):

- *The technical era*: this lasted from the beginning of public air transportation until the late 1960s. Safety occurrences were initially related to technical and technological failures, therefore safety endeavors focused in improving those areas. Efforts were successful since by the 1950s the frequency of accidents had gradually declined and safety processes had considered mainly regulatory compliance and oversight.
- *The human factors era:* this was developed from the early 1970s until the mid-1990s. Although in those days the frequency of aviation accidents was significantly reduced and aviation became a safer mode of transportation due to the investments in technology improvements, several inconceivable accidents, unrelated to systems' design failure, switched investigators' interest to human performance. Safety endeavors incorporated multiple factors having the potential to affect behavior and acknowledged that aviators operate in a complex environment.
- *The organizational era*: the organizational era began in the mid-1990s. In fact, safety thinking encompassed also the organizational factors that might potentially jeopardize safety, in addition to the already human and technical factors on board. The dominant elements that were considered were the impact of organizational culture and policies and the importance and effectiveness of safety risk controls.
- *The total system era*: this was introduced from the beginning of the century. Nowadays, both safety agencies and organizations, on either a mandatory or voluntary basis, are following a holistic approach in safety, which integrates all the elements that have been found to play an important role in safety during its evolution. In that event, they implement a safety management system (SMS), which starts from the management and is driven by the routine collection and analysis of safety data. It includes the identification

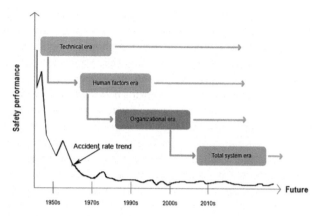

**FIGURE 1.1** Worldwide Commercial Accident Rate and Safety Thinking Evolution. (Source: Adapted from ICAO SMM 2018. Under ICAO Standard Conditions for reproduction of ICAO materials.)

of the various hazards and safety risks and provides guidance for the use of proactive strategies and methodologies, in an effort to mitigate and if possible restrict them from occurring.

A careful look at the historical statistical analysis of the worldwide commercial accident rate presented by major manufacturers (Airbus 2018; Boeing 2018) is enough to reveal the different eras of safety thinking and especially the points where the continuing downward trajectory of the curve is interrupted by an upward boom. Clearly, the excitements that caused new safety thinking were effectively treated as they were short-lived and followed by accident recession and a new downward trajectory. The whole effort should be considered satisfactory as the accident rate is kept very close to zero.

It is therefore obvious that the need to take human factors on board in the investigation reports and more recently integrate them into the safety models and training programs had been dictated by the accidents themselves. ICAO argues that although human failure can primarily be indicated as the reason for the system's breakdown, only an in-depth investigation into the hidden human factors will provide guidance as to why the breakdown occurred (2003). Therefore, in contemporary safety thinking human factors should be the starting point rather than the concluding point.

Nevertheless as previously suggested, in an area where human error remains the primary contributor in more than 70 percent of accidents (Reason 2008; Wiegmann and Shappell 2003) and where safety incidents are still identified in the mandatory occurrence reports, the need for further human factors integration should become a priority. Maintaining the accident rate at very low levels is an achievement but cannot under any circumstances be regarded as a mathematical constant. The estimations for a gradual increase of air traffic reveals that the actual number of fatal accidents and consequently the related number of fatalities have the potential to rise, without necessarily impacting the overall steady performance, or even the declining trend of the accident rate.

Focusing on aircraft maintenance, where unsafe behaviors are still being reported in combination with complaints from organisations about the delivery of low-skilled engineers, the need for human factor integration should be enhanced not only in accident investigation processes or error mitigation strategies but also in initial training where safety culture is primarily being developed.

Human failures in the worst and rare scenarios that lead to aircraft accidents have severe immediate and short- and long-term implications that are costly in terms of loss of lives and economically. Aviation disasters unquestionably disturb public trust in the industry, generate significant negative effects on investors' loyalty to the industry (Kaplanski and Levy 2008) and question the legitimacy of the airline involved to such an extent that it may force closure of the airline (e.g. the case of Helios airways that ceased ops immediately after a fatal accident in 2005).

Whenever human failures are captured in a timely manner before they develop, which is the most frequent and inconsequential scenario, they can "impose significant costs when flights are delayed or cancelled" (ICAO 2003). On these occasions they result in a series of implications such as the loss of good reputation for on-time performance and a significant increase in airline costs through a burdening in direct operating costs (Wells 1999). It is therefore self-evident that investing in human factors training and integration is not just a theoretical or legislative obligation but a practical necessity to ensure safety and survivability of the industry.

## 1.2 THE IMPORTANCE OF TRAINING IN HUMAN FACTORS

The importance of knowledge and application of the human factor is certainly demonstrated by the need that has led the industry to introduce it into the fields of initial training and aircraft maintenance and inspection. It is also supported by the results of the integration of human factors or simply from the implications that the industry experiences whenever the importance of human factors is crowded out or ignored. A report conducted by the UK's Civil Aviation Authority released in 2015 that analyzed more than 4000 reports under organizations' maintenance error management system revealed that maintenance human performance is still a causal/contributory factor in a significant number of events and further guidance is required to identify best practices in order to reduce the likelihood of errors (2015).

As previously suggested, knowledge and training in human factors is nowadays to some extent regulated or perceived as fundamental for the safe and effective execution of all aircraft-related operations and the protection of the health and the physical and mental integrity of workers. The key reasons for the importance of human factors is to mitigate human error, and are well-documented in numerous contributions (Johnson 2014; Latorella and Prabhu 2000; EASA 2012; NLR Air Transport Safety Institute 2010; ICAO 1998); they include the fostering of a positive safety culture through hazard identification and risk assessment, improved situational awareness, usage of incident reporting systems, convincing leadership and effective development of organizational safety management systems.

The need for continued alertness regarding the effects of the human factor in aircraft maintenance is endorsed also by incidents that affect the health of workers. In 2013, according to a report released in Canada by Royal Canadian Air Force

(Aerossurance 2016), an aircraft engineer was seriously injured while performing a task at the flap area of an airbus aircraft. The engineer was hit by a wing spoiler that was activated inadvertently by another technician who applied hydraulic power to the aircraft. The preliminary investigation of this incident revealed deficiencies in aircraft maintenance safety procedures and, among others, facilities, organization and personnel disruptions.

In another report released in UK, two aircraft maintenance engineers were seriously injured in 2015 while performing a task on the tail of an airplane. Both fell from the mobile elevated work platforms that were knocked over by aircraft's airbrake that was activated when another employee closed the wrong circuit breaker. The court dealing with the case fined the maintenance organization and reminded aircraft maintenance companies that "not all risks are covered by the Aircraft Maintenance Manual and additional measures need to be introduced" (SHP 2017).

## 1.3 INCIDENTS/ACCIDENTS ATTRIBUTABLE TO MAINTENANCE HUMAN FACTORS

As previously noted, the evolution of safety thinking switched from technology oriented to human factors in the outcome of the particularities of fatal accidents and serious incidents that occurred from errors in the maintenance process somehow connected to unusual or unpredictable behavior on behalf of the maintenance personnel and are briefly cited below. Further to those accidents, the need for continuous integration of human factors impact in aircraft maintenance, irrespective of the great progress achieved in all areas, and in particular the ongoing efforts to foster safety culture and develop a systematic safety management system within maintenance organizations, is undeniably revealed through recent accidents/incidents manifesting aircraft maintenance. To this end, selected accidents/incidents are cited here and in the following chapters as evidence for the need to keep human factors in the headlines and to instigate authorities to promote further research in human factors impact and training methodologies.

### 1.3.1 INCIDENT/ACCIDENT EXAMPLE 1: THE CASE OF ALOHA AIRLINES, APRIL 28, 1988

A Boeing 737–200 operated by Aloha Airlines Inc. experienced an explosive decompression and structural failure at 24,000 feet, while flying to Honolulu, Hawaii. In particular, almost 18 feet of the cabin skin and structure aft of the cabin entrance door and above the passenger floor line separated from the airplane during flight. There were 89 passengers and six crewmembers on board. One flight attendant was ejected from the fuselage during the decompression and was not found. Seven passengers and one flight attendant were seriously injured. The flight crew performed a successful emergency landing at Kahului Airport on the Island of Maui (Figure 1.2).

According to the accident report conducted by the National Transportation Safety Board (NTSB) the probable cause was the failure of the Aloha Airlines maintenance program to detect the presence of significant disbonding and fatigue damage which ultimately led to failure of the lap joint and the separation of the fuselage upper lobe.

**FIGURE 1.2** Boeing 737–200 Incident, April 28, 1988. (Source: National Transportation Safety Board. In the public domain.)

The report raised safety issues manifesting the human factors aspects of air carrier maintenance and inspection for the continuing airworthiness of transport category airplanes (1989).

Indeed, the accident investigation team identified the human factors aspects that contributed to the maintenance and inspection task associated with the accident. A prominent example of the finding was that

> a person can be motivated to do a critical task very well but when asked to perform the same task repeatedly, factors such as expectation of results, boredom, task length, isolation during the inspection task, and the environmental conditions all tend to influence the performance reliability. Another factor [...] pertains to the effect of circadian rhythms on human behavior. Airline maintenance is most often performed at night and during the early morning hours [...] the time of day documented to cause adverse human performance. Another factor that may have affected performance was [...] insufficient aircraft downtime to perform the task [...] mechanics and inspectors are forced to perform under time pressure [...] and were reluctant to keep airplanes in the hangar any longer than absolutely necessary.

Aloha's accident is considered as the reason that in 1989 the Federal Aviation Administration (FAA) sponsored a research program on human factors in aviation maintenance and produced guidance on countermeasures to mitigate human error.

### 1.3.2 Incident/Accident Example 2: The Case of British Airways, June 10, 1990

A BAC One-Eleven airplane climbing through 17,300 feet on departure from Birmingham had its left windscreen blown out. Despite the fact that the captain was sucked halfway out of the windscreen aperture and was restrained by cabin crew, the co-pilot landed the airplane safely at Southampton Airport. The passengers and crew did not suffer any serious injuries.

The investigation conducted by a team of the UK's Air Accidents Investigation Branch (AAIB 1992) found that the windscreen was blown out under the effects of cabin pressure which overcame the retention of the securing bolts; 84 of these, out of a total of 90, were of smaller than specified diameter, and the report concluded that one individual who had carried out a safety critical task failed to perform as expected and according to the manuals. To this the team examined factors such as environmental conditions, poor human performance due to circadian effects, poor judgment and working practices, inadequate access to the job, shifts, sleep deprivation, workload and the training of maintenance engineers.

### 1.3.3 INCIDENT/ACCIDENT EXAMPLE 3: THE CASE OF EXCALIBUR AIRWAYS, AUGUST 26, 1993

An A320–212 airplane, during its first flight after a flap change, exhibited an undemanded roll to the right on take-off, a condition which persisted until the aircraft, after one missed approach, landed safely back at London Gatwick Airport, under a degraded flight control system due to loss of spoiler control. The investigation conducted by AAIB identified the causal factor as a maintenance error committed during the flap change of the airplane (1994). In particular, the spoilers were left in maintenance mode following an incomplete procedure and in the absence of reinstatement and functional check of the spoilers after flap fitment.

Investigators highlighted various human factors contributing to the main causal factor such as failure of maintenance to comply with the manuals, poor supervision, lack of knowledge, inadequate briefing during the shift handovers, distraction and overload due to unscheduled maintenance.

### 1.3.4 INCIDENT/ACCIDENT EXAMPLE 4: THE CASE OF AIR MIDWEST, JANUARY 8, 2003

A Beechcraft 1900D crashed shortly after take-off from Charlotte-Douglas International Airport, Charlotte, North Carolina (Figure 1.3). The two flight crew-members and 19 passengers aboard the airplane suffered fatal injuries and one person on the ground received minor injuries, while the airplane was completely destroyed. The investigation (NTSB 2004) identified as the probable cause the airplane's loss of pitch control during take-off, which resulted from the incorrect rigging of the elevator control system after quality assurance missed a critical step that could have prevented the accident. The report deals with several human factors issues that contributed to the accidents manifesting maintenance work practices, poor procedures and documentation, lack of oversight and poor training as the overarching factors.

**FIGURE 1.3**    Crash of Beechcraft 1900D, January 8, 2003. (Source: Aviation Safety Net. Photographer: Lookout2. Under Creative Commons Attribution-Share Alike 3.0 License.)

### 1.3.5 Incident/Accident Example 5: The Case of China Airlines (Taiwan), August 20, 2007

A Boeing 737–800 landed at Naha Airport as planned and taxied to the apron. At the stand, fuel gushing from an area near the no. 2 engine pylon ignited. After all occupants had safely evacuated, a large explosion occurred in the center of the airplane and caused a fire, which left only part of the airframe intact (figure 1.4). The investigation report conducted by the Japan Transport Safety Board identified a causal chain that contributed to the accident (2009). It was considered as highly probable that when pilots retracted the slats after landing, the track can that housed the inboard main track of a slat surface on the right wing was punctured, creating a hole. As a result, fuel that leaked out came into contact with high-temperature areas of the engine and exploded. Investigators identified as the cause of the puncture in the track can, the detachment of the downstop assembly that was incorrectly maintained about one and a half months prior to the accident.

Based on the report, among other factors, poor planning and implementation of maintenance task, inadequate supervision, the improper design of the downstop assembly and inadequate physical access had been the human factors issues contributing to the accident. Despite the fact that the downstop assembly was in a location difficult to access neither the manufacturer nor the airline paid sufficient attention to this, while maintenance had not reported the related difficulties of the task.

### 1.3.6 Incident/Accident Example 6: The Case of British Airways, May 24, 2013

An Airbus A319, during take-off from London Heathrow airport, experienced damage on the airframe and a number of aircraft systems and an external fire development from leaking fuel on the right engine, due to the detachment of the fan cowl

**FIGURE 1.4** Boeing 737–809 Accident, August 20, 2007. (Source: Aviation Safety Net. Photographer: Thomas Mitchell. Under Creative Commons Attribution-Share Alike 3.0 License.)

doors from both engines. The pilots performed a single-engine emergency landing at the airport and manage to evacuate all passengers safely.

Investigators (AAIB 2015) identified as a causal factor for the accident a maintenance error which had occurred in a scheduled maintenance on the night previous to the accident that had led to the fan cowls on both engines being left unlatched. Many human factor issues are elaborated in the report such as fatigue, night shift working, poor planning, inadequate visual inspection and aircraft swap error and non-compliance with the procedures of the manual.

**Lessons Learned #1**

The importance of attaining knowledge in human factors, as well as the need to continuously integrate this expertise into error-mitigation training by all aircraft maintenance engineers, is unquestionably justified by the above events. The role of human factors in specific actions that jeopardize safety is often hidden from the direct causal factors and will remain latent unless special attention is given to this issue.

## 1.4 MURPHY'S LAW APPLICATION IN MAINTENANCE

Errors can be made by all individuals involved in a particular task. Complacency is one of the most important reasons that drives even the best trained and the most skilled and experienced personnel to develop behaviors contrary to what was expected by entering a state of self-satisfaction accompanied by a loss of awareness of the dangers. This situation is reminiscent of Edward A. Murphy's "Law" that actually refers to safety-critical system aiming for defensive design. The law, that is often misunderstood, states that "anything that can go wrong will go wrong," (Figure 1.5) which in maintenance has the meaning that if there are more than two ways to the one correct to assemble a component or install a component to the system, then it is very likely that someone will follow the wrong way. Research has shown that as time passes, familiarity breeds complacency (Peters and Peters 2006), and as a result people that are doing familiar and repetitive work are no longer engaged in the task (Salas and Maurino 2010). Therefore, Murphy's Law must be always recalled by aircraft maintenance engineers who perform repetitive tasks in complex systems such as aircraft inspection.

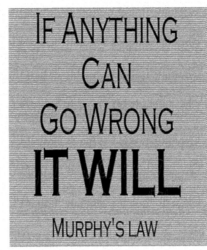

Thinks shortcuts save time

Puts away manuals as experience grows

Overlooks bad norms and unsafe behavior

Feels complacency enhances performance

**FIGURE 1.5**    Murphy's Law. (Source: Created by Yiannakides and Sergiou 2019.)

## REFERENCES

AAIB (Air Accidents Investigation Branch). UK Department of Transport. *Report No. AAR 1/92*, released February 1992. Retrieved from Aviation Safety Network. Flight Safety Foundation.

AAIB (Air Accidents Investigation Branch). UK Department of Transport. *Report No. AAR 2/95*, released December 16, 1994.

AAIB (Air Accidents Investigation Branch). UK Department of Transport. *Report No. AAR 1/2015*, published July 14, 2015. Retrieved from Aviation Safety Network. Flight Safety Foundation.

Aerossurance. November 14, 2016. *Serious Injury during Aircraft Maintenance*. http://aerossurance.com/safety-management/injury-aircraft-maintenance.

Airbus. 2017. *Global Market Forecast. Growing Horizons 2017/2036*. Airbus SAS.

Airbus. 2018. *A Statistical Analysis of Commercial Aviation Accidents 1958–2017*. Airbus SAS.

Bainbridge, L. 1999. Process Underlying Human Performance. In *Handbook of Aviation Human Factors*, ed. Daniel J. Garland, John A. Wise, and V. David Hopkin, 159. Mahwah, N.J.: L. Erlbaum Associates Publishers.

Boeing. 2013. *Maintenance Error Decision Aid (MEDA), User's Guide*.

Boeing. 2017a. *Pilot and Technician Outlook: 2017–2036*. www.boeing.com/commercial/market/pilot-technician-outlook/2017-technician-outlook.

Boeing. 2017b. *Current Market Outlook 2017–2036*. Seattle: Boeing Commercial Airplanes.

Boeing. 2018. *Statistical Summary of Commercial Jet Airplane Accidents. Worldwide Operations 1959–2017*. Seattle: Product Safety Organization, Boeing Commercial Airplanes.

CAA (Civil Aviation Authority) UK. 2015. *Aircraft Maintenance Incident Analysis*. CAP 1367.

CASA (Civil Aviation Safety Authority) Australia. 1998. *Regulation 66.015/Part66 Manual of Standards*.

CASA (Civil Aviation Safety Authority) Australia. 2013. *Human Factors. Resource Guide for Engineers*.

Choi, Hoon-Seok, and John M. Levine. 2004. Minority Influence in Work Teams: The Impact of Newcomers. *Journal of Experimental Social Psychology*. 40(2): 273–280.

Civil Aviation Administration of China. 2005. *Regulations on Certification of Civil Aircraft: Maintenance Personnel Training Organizations.* CAAC Decree No. 154.

Commission Regulation (EU) No. 1321/2014 of November 26, 2014 on the continuing airworthiness of aircraft and aeronautical products, parts and appliances, and on the approval of organisations and personnel involved in these tasks.

Dempsey, P. S., and Lawrence E. Gesell. 1997. *Air Transportation: Foundations for the 21st Century.* Arizona: Coast Aire Publications.

Depaolo, Concetta, and Constance H. Mclaren. 2006. The Relationship Between Attitudes and Performance in Business Calculus. *INFORMS Transactions on Education.* 6(2): 8–22.

EASA (European Aviation Safety Agency). 2012. *European Strategy for Human Factors in Aviation.*

European Parliament. 2009. *The Impact of the Economic Crisis on the EU Air Transport Sector.*

FAA (Federal Aviation Administration). 2014. *Human Factors in Aviation Maintenance, Operator's Manual.*

FAA (Federal Aviation Administration). 2017. *Maintenance Human Factors Training.* Advisory Circular 120–72A.

Goyal, R., and D. Negi. 2014. Impact of Global Economic Crisis on Airline Industry. *International Journal of Commerce, Business and Management* (IJCBM). 3(2): 297–301.

HSE (Health and Safety Executive). 1993. *ACSNI Human Factors Study Group: Third report – Organising for safety.*

IATA (International Air Transport Association). 2018a. *Annual Review 2018.*

IATA (International Air Transport Association). 2018b. *Future of the Airline Industry 2035.*

ICAO (International Civil Aviation Organization). 1998. *Human Factors Training Manual.* Doc. 9683.

ICAO (International Civil Aviation Organization). 2003. *Human Factors Guidelines for Aircraft Maintenance Manual.* Doc. 9824.

ICAO (International Civil Aviation Organization). November 29, 2017. NGAP Summit Addresses Pressing Shortages of Skilled Professionals for Future Air Transport Network. www.icao.int/Newsroom/Pages/ICAO-NGAP-Summit-addresses-pressing-shortages-of-skilled-professionals-for-future-air-transport-network.aspx.

ICAO (International Civil Aviation Organization). 2018. *Safety Management Manual.* 4th ed. Doc. 9859.

Japan Transport Safety Board. Report No. AA2009-7, released August 28, 2009. Retrieved from Aviation Safety Network. Flight Safety Foundation.

Johnson, B. 2018. Training Solutions to Meet the Looming Maintenance Personnel Shortage. *Aviation Mx. Human Factors.* 6: 3–5.

Johnson, W. B. 2014. Human Factors Training. Evolution and Reinforcement. In *Operator's Manual for Human Factors in Aviation Maintenance.* Federal Aviation Administration.

Kaplanski, G., and H. Levy. 2008. Sentiment and Stock Prices: The Case of Aviation Disasters. *Journal of Financial Economics.* 95(2): 174–201; revised 2010.

Karlsson, Torbjorn. 2007. Survey reveals skills shortage in airline engineering. *Flight Airline Engineering.* www.flightglobal.com/news/articles/survey-reveals-skills-shortage-in-airline-engineerin-217000.

Latorella, K. A., and P. V. Prabhu. 2000. A Review of Human Error in Aviation Maintenance and Inspection. *International Journal of Industrial Ergonomics.* 26(2): 133–161.

Lewis, Paul A. 2012. Flying High? A Study of Technician Duties, Skills, and Training in the UK Aerospace Industry. *SSRN Electronic Journal.*

Macário, R., V. Reis, and C. Marques. 2009. *Scenarios for Air Transport, AirNets.* MIT Portugal program.

Ministry of Civil Aviation of India. 2016. *Standard Apprenticeship for Aircraft Maintenance Engineers in Civil Aviation Sector.*

NLR Air Transport Safety Institute. 2010. *Aircraft Ground Handling and Human Factors.*

NTSB (National Transportation Safety Board). Report No. AAR-04/01, released February 26, 2004. Retrieved from Aviation Safety Network. Flight Safety Foundation.

NTSB (National Transportation Safety Board). Report No. AAR-89/03, released June 14, 1989. Retrieved from Aviation Safety Network. Flight Safety Foundation.

Patterson, Ian R. 2002. *Organisational Culture and Safety Culture as Determinants of Error and Safety Levels in Aviation Maintenance Organisations: A Latent Failure Approach.* Thesis submitted in partial fulfilment of the requirements for the degree of PhD in Psychology at Massey University, Albany, New Zealand.

Peters, George A., and Barbara J. Peters. 2006. *Human Error: Causes and Control.* Boca Raton, Fla: CRC Press.

Reason, J. 2008. *The Human Contribution: Unsafe Acts, Accidents and Heroic Recoveries.* Boca Raton, Fla: CRC Press.

Salas, Eduardo and Dan Maurino (eds.). 2010. *Human Factors in Aviation.* Burlington, Mass: Academic Press.

SHP (Safety and Health Practitioner). March 17, 2017. Court Hears how Workers Fell from the Tail of a Plane. www.shponline.co.uk/in-court/court-hears-how-workers-fell-from-plane.

US Government Accountability Office. February 28, 2018. *Aviation Workforce: Current and Future Availability of Aviation Engineering and Maintenance Professionals.* GAO-14-237.

Wells, A. T. 1999. *Air Transportation. A Management Perspective.* Belmont, Calif.: Wadsworth.

Wiegmann, D. A., and S. A. Shappell. 2003. *A Human Error Approach to Aviation Accident Analysis: The Human Factors Analysis and Classification System.* Aldershot, UK: Ashgate.

# 2 Human Performance and Limitations

When examining human factors we should be able to understand the human sensory system and the way that our brain receives and processes all the incoming information, in order to respond to external and internal stimuli. The goal of course is not to acquire in-depth knowledge on medical or psychology issues, but rather to understand the associated limitations and their impact on human performance that generate error potential. Consequently, the following section reviews the functions of vision and hearing as well as the sequence of information processing. Likewise the involvement of the senses of attention and perception and the contribution of different types of memory are focused upon with a view to delineate the various choices in decision-making. Wherever appropriate, these deficiencies are associated with the daily tasks in aircraft maintenance, with a particular focus on the functions performed in conditions, where the human sensory system becomes less capable, such as work in confined space areas and at significant heights.

## 2.1 VISION

It is much easier to understand the process of our vision if we compare it to the function of a common camera (American Academy of Ophthalmology 2013). The components of the camera contribute to the process in a similar manner as the parts of our eye. Thus, as a camera has a focusing glass to collect the light and the aperture to control the amount of light that will enter the device, the same occurs in our eye, where the cornea acts as the focusing device and the pupil controls the amount of light with the help of the iris. In comparison to in the camera where light falls on the film and the processing of the image begins, in the case of our vision, the visual image begins to form when the light rays that are reflected by an object fall on the retina, after initially being focused by the cornea and the lens. At that point the incoming light stimulates the rods and cones that are light-sensitive cells on the retina, thus generating small electrical impulses, which travel through the optic nerve to the visual cortex in the brain. Our brain concludes the process of vision by interpreting the electrical impulses and perceiving images.

The various parts of our eye are depicted in Figure 2.1 and their functioning (Lang 2000) could be summarized as follows:

- The *cornea* incorporates the focusing ability of our eye for 70 percent up to 80 percent. It is located at the very front of the eye and has the form of a clear window.
- The *iris* is the colored thin structure of the eye that monitors the quantity of light allowed to enter the eye by controlling the size of the pupil.
- The *pupil* is a dark opening inside the iris, which adjusts its size according to the different light levels, thus allowing light to enter the eye. It has the ability to change its size in diameter from 1.5 mm to more than 8 mm.
- The *lens* is located behind the iris and further assists the focusing process of incoming light.
- The *retina* is located on the rear wall of the eyeball and is a layer of nerve cells that are connected to the optic nerve. It senses light, and creates electrical impulses that travel through the optic nerve to the brain.
- The *optic nerve* transfers the image in the form of electrical impulses from the retina to the brain.
- The *rods and cones* are two types of light sensitive cells that are located on the retina. Cones are active at higher light levels (photopic vision) and are capable of achieving color vision and high spatial acuity. Cones are mainly located in the central area of the retina, which is known as the fovea and assist human vision to distinguish about 1000 different shades of color. As we move outwards, the cones become less dense and are progressively replaced by rods, which are concentrated in the periphery of the retina. Rods are responsible for vision at low light levels (scotopic vision) since they are much more sensitive in that area. They do not mediate color

# HUMAN EYE ANATOMY

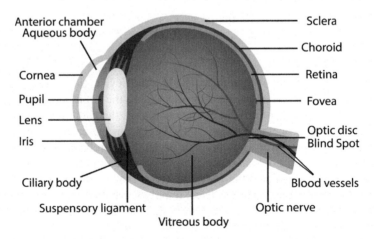

**FIGURE 2.1**    Human Eye Anatomy. (Source: AdobeStock. © Neokryuger-stock.adobe.com. Under Standard License AdobeStock_105281811.)

vision and have a low spatial acuity. They are poor at distinguishing fine detail, but excellent at detecting movement at the edge of the visual field (peripheral vision). As light decreases and we approach lower light levels, the sensing task is passed from the cones to the rods, whilst the resulting image forms only in black and white and shades of grey.

## 2.1.1 Factors Affecting Human Vision

The most common ophthalmological disorders affecting human vision are as follows:

- *Myopia* is the eye disorder in which the light rays are not concentrated in the retina but at some point in front of it. If not corrected, it causes blurred perception of distant objects. Correction is achieved with eyeglasses, contact lenses or laser surgery.
- *Astigmatism* is the condition in which it is not possible to gather incoming light rays at one point, thus resulting in distant and nearby objects being disfigured. The correction is achieved in a similar manner to myopia with eyeglasses, contact lenses or laser surgery.
- *Hyperopia* is the condition in which a parallel beam of rays does not focus on the retina but beyond, so the eye is capable of distinguishing neither distant nor nearby objects. Correction is done as above with eyeglasses, contact lenses or laser surgery.
- *Presbyopia* is a condition that reduces vision for nearby objects and causes fatigue. The majority of people experience this condition in the fifth decade of life and similar to the aforementioned disorders, it can be corrected with eyeglasses, contact lenses or laser surgery.
- The *cataract* is a condition of the eye that occurs due to normal aging of the body and causes a gradual reduction in vision. It can be corrected surgically.

Further to the above physical disorders, smoking and the use of alcohol and narcotic drugs can lead to further complications such as blurry vision. In addition, poor lighting and the low quality of the surrounding environment may downgrade vision clarity. It is noteworthy that when we move from a bright environment to a darker one, a short span of time is necessary for our vision to adjust. The ability of the eye to adapt itself to decreasing illumination is called dark adaptation and occurs after 30 minutes of dark adaptation time (Khurana 2007).

Finally, a condition that aircraft engineers must be aware of, especially when they perform close visual inspection is the *blind spot* (Figure 2.2), which is not a physical disorder of human vision but a constructional anomaly that all people experience. It is the situation where at a certain distance, various details of an image cannot be perceived, depending of course on the specific size of the image in question. This is due to the absence of photoreceptors at the point where the optic nerve attaches to the retina and the inability of the eye to create electrical impulses. This point is slightly to the right of the right eye and slightly to the left of the left eye and

**FIGURE 2.2**    Eye Blind Spot. (Source: Created by Yiannakides, December 24, 2018.)

represents a void in the field of vision. We do not perceive this void because the brain fills it with data from the surrounding area and information collected from each eye separately. In simple words, an image falling at this point will not be detected. One method of dealing with the blind spot is to perform visual scanning techniques with frequent eye movement and by reducing the time we look in any one direction.

### 2.1.2 The Sense of Vision in Aircraft Maintenance

Unlike pilots who are required via regulations to maintain excellent vision and are medically examined on an annual basis, engineers are not subject to this require-ment. However, they are usually obliged to undergo an initial medical examination regarding their ability to distinguish colors (color defective vision), which is usually imposed by employers. Notwithstanding the fact that a comparable legal obligation has not yet been introduced, the nature of the work undertaken by aircraft engineers requires excellent vision, since tasks are usually carried out in all kinds of environ-mental and lighting conditions and are thus visually demanding, especially those related to aircraft inspection and avionics.

Unquestionably, the development of technologies has enhanced the ability to detect any kind of abnormalities on the aircraft's surfaces, systems or wiring, but the need for initial visual detection and repair by the engineer still exists. The NTSB has been very strict on this issue by repeatedly underlining that accidents have been caused due to the inability of inspectors to detect corrosions that could have been visually identified, while the FAA insists that the process of inspection requires a high level of visual acuity (2014). Of course, impaired vision is not the absolute reason behind aircraft inspection failures, but other human factors also contribute. Consequently, it is imperative that aircraft engineers comprehend the need to maintain their vision in an excellent condition and to take all the necessary precautions or corrective measures. For example, the CAA UK advises engineers who wear contact lenses and work in hangars that are dry or dusty, to avoid wearing their lenses for periods greater than 12 hours, otherwise they may experience eye irritation (2002).

**Lessons Learned #2**
Aircraft maintenance tasks are performed in all kind of environmental conditions and are visually demanding therefore vision should be examined regularly.

## 2.2 HEARING

In order to effectively comprehend the impact of noise on the sense of human hearing in the workplace, it is essential to gain knowledge of the processes of sound and hearing. Thus, sound could be conceived as pressure waves generated from a vibrating object and propagated through a medium due to changes in atmospheric pressures. Sound is perceived by the human sensory organ, which is simply the ear, a special receptor organ structured to respond to these sound waves. The frequency (pitch) of sound measured in hertz (Hz) and its loudness (volume/intensity) measured in decibels (dB) are among the basic characteristics of sound. The human ear can typically perceive sounds in the frequency range of 20 Hz to 20 kHz.

The perception of sound (Everest 2001; Ballou 2015; Moore 2013) starts with the vibration of the eardrum inside the ear, which is caused by the sound waves that reach the pinna. The sound is then transmitted in the form of an acoustic pulse through the auditory canal to the ossicles that are located in the middle ear. The ossicles amplify the sound waves and transfer them to the inner ear and the cochlea. The fluid within the cochlea is then set in motion and captured by the auditory nerve that transfers the sound in the form of nerve impulses to the brain, in order to interpret the sound.

Hearing is not the only function undertaken by the human ear. The inner ear and in particular the vestibular system is responsible for maintaining the body's postural equilibrium, which is a state of balance. In brief, three semicircular canals are filled with a fluid and are arranged in three planes, 90° to each other. The semicircular canals respond to rotational movements pertaining to angular acceleration, while other organs within the vestibule react to changes in the position of the head with respect to gravity and linear acceleration. The state of balance can be affected by head injuries, eating disorders, illness and drug use, as well as alcohol consumption that can persist for up to three days following use.

In particular, the main parts of the ear are depicted in Figure 2.3 and include the following:

- The *outer ear*, which consists of the pinna and the external auditory canal.
- The *tympanic membrane* or *eardrum* that is located between the outer and the middle ear.

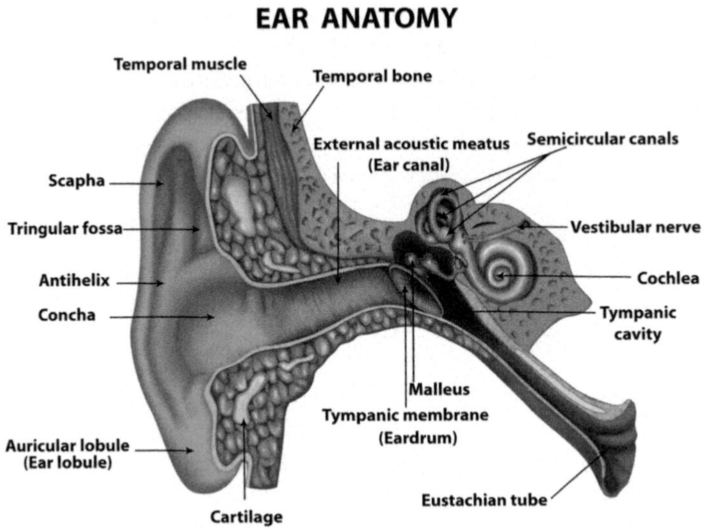

## EAR ANATOMY

**FIGURE 2.3** Anatomy of the Ear. (Source: AdobeStock. ©vishalgokulwale-stock.adobe. com. Under Standard License AdobeStock_203675370.)

- The *middle ear,* which consists of the ossicles (malleus, incus and the stapes) that transmit the sound waves to the inner ear and the Eustachian tube that assists in equalizing the pressure in the middle ear, a prerequisite for the transfer of sound waves.
- The *inner ear,* which consists of the cochlea that contains the nerves for hearing and the vestibule and semicircular canals that contain the receptors for balance.

### 2.2.1 THE IMPACT OF NOISE EXPOSURE

Research has revealed that aircraft maintenance workers are exposed to hazardous noise levels on a daily basis that are above the recommended regulations as concerns industrial and work environments, with an average volume of 80 dB that sometimes reaches a peak value of 119<db>dB (Smedje et al. 2011; Noweir and Zytoon 2013). These elevated noise levels are prevalent in aircraft maintenance contexts like hangars and aprons and are generated from many sources, such as aircraft and vehicle movements, engine test runs and pneumatic tool operations. Inevitably, they constitute a constant burden on the health of workers and in many cases have the potential to contribute to precarious actions and behavior. The FAA suggests that exposure to sound may have both physiological and psychological effects (1998) and notes the importance of remembering that the duration of exposure to high-intensity sounds determines the potential risk.

Beginning with the physiological effect, researchers note that for a sound with a volume of approximately 120 dB and a pitch between 500 Hz and 10 kHz, people begin to experience discomfort. If density is increased above 140 dB the threshold of pain is reached, while a few minutes of continuous exposure to noise above 140 dB can result in permanent damage to various elements of the ear and hearing loss (Barron 2003). Any damages to the ossicles or the eardrum could result in a degradation of hearing known as *conductive deafness,* while any damage to the hair cells in the inner ear, which is usually not immediately noticeable, can cause *noise induced hearing loss* (NIHL) (Allen 2013). Another hearing impairment is *hearing loss with aging* (presbycusis), which involves the loss of hearing that gradually occurs in most people as they age.

Concurrently, exposure to noise can generate psychological effects that have the potential to impact overall health and safety performance, since noisy environments may generate *stress, fatigue, communication failures* and *distraction.* To this end, the European Parliament and the Council, through its Directive 2003/10/EC "On the minimum health and safety requirements regarding the exposure of workers to the risks arising from physical agents (noise)," defines the exposure limit values and exposure action values in respect of the daily noise exposure levels and peak sound pressure (2003). In general, if workers do not take precautions (ear plugs/headsets) on a daily basis, permanent hearing loss could occur, following unprotected exposure to loud noise (higher than 87 dB) for eight or more hours per day for several years, while eardrum rupture may occur during exposure at 140 dB peak noise level.

According to the same regulation, when carrying out their noise risk assessment, employers shall give particular attention to

The level, type and duration of exposure, including any exposure to impulsive noise; [...] any effects concerning the health and safety of workers belonging to particularly sensitive risk groups and any indirect effects on workers' health and safety resulting from interactions between noise and warning signals or other sounds that need to be observed in order to reduce the risk of accidents.

Maximum recommended noise dose exposure limits vary from country to country, but it is important to note that they all converge on the view that these limits should be lower in the case of 12-hour shifts compared to those for eight-hour shifts, a fact which should be taken seriously by both the management and aircraft engineers. Therefore, if it is considered that the lower limit for which protection is required is also the safest, then the resulting conclusion is that the recommended point for hearing protection is 85 dB.

### Lessons Learned #3

Aircraft hangars and aprons are very noisy environments. Protect your hearing from hazards and safeguard yourself and aircraft from potential unsafe acts. Use hearing protection equipment such as earplugs, earmuffs and simple or communication headsets whenever sound exceeds the permissible noise levels, having in mind that due consideration should be given to attaining appropriate situational awareness of the task.

**TABLE 2.1**
**Limits for Permissible Noise Exposure**

| Duration | Maximum Recommended Noise Dose Exposure |
|---|---|
| 8 hours | 90 db |
| 6 hours | 92 db |
| 4 hours | 95 db |
| 3 hours | 97 db |
| 2 hours | 100 db |
| 1.5 hours | 102 db |
| 1 hour | 105 db |
| 30 minutes | 110 db |
| 15 minutes | 115 db |

*Source:* US Department of Labor. In the Public Domain.

## 2.3 INFORMATION PROCESSING

Knowledge of the human brain's process when dealing with incoming information and decision-making is essential, since it entails a psychological process involving human performance through interaction with systems. Information processing refers to the human ability to receive, interpret and respond within a given timeframe to external stimuli. Although research may be ongoing, there is a convergence of views that the human brain processes the information it receives by following a series of steps and by applying our senses and mental capabilities. Wickens et al. in the book *Engineering Psychology and Human Performance* present a model of information processing (Figure 2.4) where *sensing, attention, memory and perception, decision-making, response* and *feedback* evolve in series and simultaneously (2016). Martinussen and Hunter suggest that the model developed by Wickens and Holland is one among the most influential, most probably because it is a useful tool for understanding and predicting human interaction with complex systems (2018).

Therefore, logically, one can assume that the way our brain processes information is very similar to the functioning principles of computers, since it incorporates input and processing, storage and output. Considering the potential impact of emerging artificial intelligence technology and its gradual and widespread future application in computers, it is no exaggeration to assume that human and computer information processing will eventually feature much more similarities than ever before.

The mental processes that participate in the Wickens and Holland model are shown in Figure 2.4. They all fall in the field of *cognitive psychology* and are described in brief in the following pages.

### 2.3.1 ATTENTION AND PERCEPTION

As already described, our vision and hearing senses act as the input gate of all stimuli around us. Our ability to focus on different stimuli is called *attention*, which in

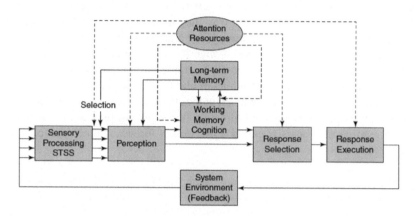

**FIGURE 2.4** Model of Human Information Processing. (Source: Christopher D.Wickens, Justin G. Hollands, Simon Banbury, and Raja Parasuraman. 2016. *Engineering Psychology and Human Performance*. Psychology Press. With permission.)

psychology can be determined as the process involving the concentration of thinking (cognitive processes).

Although it has not yet been clarified how *attention* functions, modern studies (Sternberg and Sternberg 2012; Goldstein 2011; Solso 2001) suggest that the initiation of information processing requires a stimulus, as well as a conscious choice to concentrate our attention on the stimulus for a specific time through the available but limited cognitive resources. By examining the modern view, we could distinguish attention into at least two basic functions. The first involves *selective attention,* where we choose to follow a certain stimulus more, at the expense of all other stimuli that compete to gain our attention and which exist at the same time in our visual or auditory field. Often, in the context of selective attention, the term "cocktail party effect" is used, which is nothing more than the ability we have to shift our attention from one stimulus to another, without completely abandoning the first. Second, *divided attention* concerns the situation where we choose to distribute our attention among two or more stimuli at the same time. However, it is noteworthy that this option involves limitations concerning the intended level of perception. It is important to note that strict separation of the two functions is difficult as they continuously vary in terms of concentration levels and time allocation. A characteristic example of selective attention is the concentration of our hearing on the news bulletin whilst ignoring other surrounding noises and conversations. On the other hand, a basic example of divided attention is the ability we have to hear and perceive the news bulletin, whilst we are editing a letter on our computer. Nowadays, a third function is being introduced labelled *simultaneous attention,* where the time of concentration is prolonged to different stimuli without affecting the quality of perception.

*Perception* on the other hand is the core stage within the sequence of information processing, where we acknowledge and interpret the attained information as something meaningful. To a large extent, all people construe incoming information in the same manner and react in the way they have been taught since their childhood, based on their accumulated knowledge and experience, such as the interpretation of colors or sounds. However, congested or falsified stimuli may lead to a different interpretation and consequently to contrasting decisions and actions. This is due to the weaknesses and limitations of human sensory organs, and mainly due to the sense of vision, especially when it comes to the perception of the depth or size of the objects or the existence and direction of motion. In psychology the cause of misinterpretation of objects is called a *perceptual illusion*. Some examples follow below:

- The *geometrical-optical illusions*, such as the Ponzo illusion depicted in Figure 2.5 (Mario Ponzo, 1882–1960) or Müller–Lyer illusion (1857–1916) depicted in Figure 2.6, that demonstrate how our mind misinterprets an object's size based on its background and depth or angular interference.

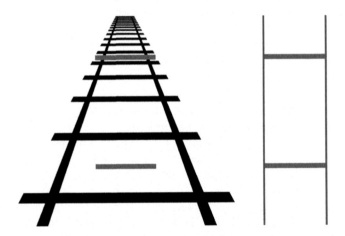

**FIGURE 2.5**  Ponzo Illusion. (Source: AdobeStock. © Peter Hermes Furian-stock.adobe. com. Under Standard License AdobeStock_212357670.)

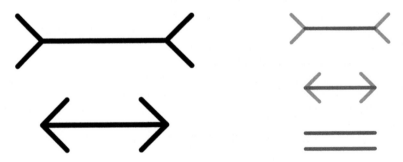

**FIGURE 2.6**  Müller-Lyer Illusion. (Source: AdobeStock. © Peter Hermes Furian-stock. adobe.com. Under Standard License AdobeStock_116545218.)

**FIGURE 2.7** Rotating Wheel Illusion. (Source: AdobeStock. © Mark J. Grenier-stock. adobe.com. Under Standard License AdobeStock_26364585.)

- The *motion illusions* (Figure 2.7) in which steady objects appear to be moving, such as the rotating wheels illusions.

Concluding, it would be an omission not to mention the *wagon-wheel effect* which, if encountered in the aircraft maintenance environment, can cause serious personal injury. This phenomenon relates to the lack of brain capacity to perceive the actual direction and speed of rotation of a grid wheel, due to visual limitations. As a result of this optical illusion, the wheel appears either to spin at a much lower angular speed or not to rotate at all. Unfortunately, the occurrence of this phenomenon in aircraft aprons and hangars has been fatal on many occasions, since rotating propellers and tail rotor blades appeared steady, when actually in full-speed rotation. In an effort to eliminate the hazards generated from the *wagon-wheel effect*, manufacturers paint the propellers and the tail rotor blades, which results in a positive illusion of a colored rotating disc, thus improving visibility.

### 2.3.2 Memory

Sternberg describes memory as "the means by which we draw on our past experiences in order to use this information in the present" (1999). It is the process of encoding, storing and retrieving information at a later time, from the original information received. The information is encoded and stored in the form of image, audio or semantics; the degree of memory capacity determines the procedure of information processing to a significant extent. One would say that the process of memory is bi-directional because it is fed according to the level of sensing and attention that the observer has made available while contributing to *decision-making* by feeding the *perception* phase.

The current literature converges to the fact that there are at least three kinds of memory, which are *sensory, short-term and long-term memory*. It is the choice of the person to store the information or alternatively to transfer it from one storehouse to another, but it seems that the success of the selection depends on the level of sensing and the duration of *attention,* while the ability to retrieve detailed information is improved by rehearsing.

Researchers suggest that *sensory memory* is very limited both in capacity and ability to retain information, since it lasts for up to four seconds for auditory stimuli and only one second for visual stimuli. On the other hand, *short-term memory* refers to the ability to store information for up to 20 seconds and is more relevant to everyday life since it builds new information over the past knowledge. *Short-term memory* is sometimes analyzed or referred to as *working memory*, which has to do with the information that we store for skills development, in order to handle tasks in a motor way, such as the driving of our car or the use of our mobile phone. Finally, *long-term memory* is the storage component where past experiences and knowledge are kept and can be retrieved decades later, blurry in their visual aspect but quite detailed in their meaning. *Long-term memory* can be distinguished into *episodic* and *semantic* memory, where the first retains the "episodes" of our life such as times and places and the latter retains the general meaning of knowledge (Tulving 1985). The information stored in episodic memory experiences a gradual deterioration in detail and accuracy over time and is finally transferred to semantic memory. The manner in which long-term memory interacts with short-term memory remains a subject of research.

Human memory is not perfect and has no infinite storage capabilities. In particular, any information burden on working memory, which is primarily used in aircraft maintenance, may lead to destruction and possible errors, since continuous concentration is required.

### 2.3.3 Decision-Making, Response and Feedback

It is reasonable to conclude that information processing would have no meaning, if it didn't lead to a specific purpose. Thus, in this case the aim is to respond to the incoming stimulus, via a task that is formed through the *decision-making* stage. From the research to date, it is concluded that the stages of *attention* and *perception,* as well as the previous knowledge stored in memory, form the *decision-making* process and determine the selection of the most appropriate choice, through different options and the deriving *course of action*. Within the *decision-making* stage, the process of *situation assessment* and reasoning also evolve, which will eventually determine whether the situation has been satisfactorily perceived or not and the weighted value of one choice over the other.

In the aircraft maintenance environment and in particular within daily complex situations that need *problem solving*, one would say that *situation assessment* is critical to indicate and acknowledge errors, since it is necessary for taking the correct course of action in the existing situation (Orasanu and Fischer 1997). The whole process is completed with the memory *feedback* stage, following the assessment of the decision taken and the action performed and depending on the result achieved, which in the case of aircraft maintenance should be the one sought without degradation and deviations.

### 2.3.4 SITUATIONAL AWARENESS IN AIRCRAFT MAINTENANCE

The term "situational awareness" is widely used in aviation and has significant deviations in interpretation with regard to the simple perception of the situation, since it includes not only the comprehension of working environment information at a given time and volume, but also the prediction of the exact status of the environment in an extended time and volume (Endsley 1995). It may seem complicated but it is a unique skill that is taught and developed rapidly in extremely dynamic environments through the daily routines of aviation.

Enhanced *situational awareness* is critical in aircraft maintenance because many decisions involve risk assessment and refer to complex problems unfolding under stressors. However, the information available for decision-making and error in aircraft maintenance is limited, as compared to the corresponding information available for pilot error in decision-making (Wiegmann and Shappell 2003). At the same time, it is unfortunate to observe that technology improvements and automation in aircraft systems has resulted in a gradual loss of situational awareness and degradation of skills, thus increasing decision errors. Any failure to integrate human factors in information processing may increase the risk of hazardous incidents, especially when dealing with complex tasks and operating under certain conditions with stressor factors. The entire information processing sequence can be affected by various factors that are either imposed from third parties and the surrounding environment or generated from us. Time pressure and distraction, poor lighting or bad weather and lack of knowledge or inadequate training are some of the factors that may lead to defective information processing and incorrect/unsafe course of actions.

### 2.3.5 INCIDENT/ACCIDENT EXAMPLE 7: THE CASE OF QANTAS AIRWAYS, OCTOBER 14, 2017

The case of the collision of a catering vehicle and a Boeing 737 is a good example demonstrating that environmental factors limiting vision and hearing can have a critical effect on engineers' situational awareness. The investigation conducted by the Australian Transport Safety Bureau revealed the sequence of events that led to the collision on the apron of Sydney airport (2018). According to the report, following a battery problem on a Boeing 737, another aircraft of the same type (VZZ) was assigned to perform the flight from Sydney to Melbourne. The aircraft was ready for departure with the pushback vehicle's towbar connected, but had to wait for the catering to be transferred from the originally assigned aircraft. The dispatch engineer discussed the aircraft status with the other engineer who had been in attendance. At the same time it had started to rain and the catering vehicle was loading the front galley and the rear on the right side.

The engineer completed his walk around the aircraft and, after noticing that the catering crew was standing close to the main entry door that was closed, he moved to the front of the aircraft and contacted the flight crew through his headset, who confirmed departure readiness. Still in rain conditions, the air traffic control cleared the aircraft for pushback and the engineer leaned out of the pushback vehicle cabin and performed a visual check to the left and right to check for vehicles. He then cleared the pushback vehicle driver to commence pushback and the flight crew to start the right engine. At that moment, the catering vehicle still behind the right wing began reversing. Although the engineer anticipated the collision and informed both the pushback vehicle driver and the crew to apply the brakes, which they did, the pins in the towbar sheared and separated from the aircraft. As a result, the aircraft collided with the catering vehicle and suffered substantial damages to the right wing surface.

The investigator revealed a loss of *situational awareness* by concluding that the engineer

> assumed that the catering truck was clear and did not visually confirm it [...] this assumption was based on an incorrect expectation of the time required for the truck to move clear [...] the engineer's view of the truck from the pushback tug's cabin was largely obstructed by the aircraft [...]. The dark and rainy conditions also made it difficult to see the truck.

## 2.4 CLAUSTROPHOBIA, ACROPHOBIA AND PHYSICAL ACCESS

*Claustrophobia* is the situation in which an individual suffers from extreme or irrational fear of confined places, whilst confined space as shown in Figure 2.8 is one which by design has limited openings for entry and exit (Bowling et al. 2002). On the other hand, *acrophobia* is a psychological disorder that emerges when exposed to heights. Both conditions can cause serious psychological effects such as severe stress and fatigue, which are capable of causing disorientation and confusion. They also have the potential to disturb information processing and *decision-making*. Since many maintenance tasks require work in the aforementioned uncomfortable conditions, such as on platforms to inspect the wings or inside fuel tanks to identify corrosion and cracks, it is a prerequisite for engineers who undertake these types of tasks to be skilled, confident and fully aware of the psychological and physical factors that may compromise their performance.

Maintenance tasks in confined spaces are accompanied with a high error probability, while working at significant heights has been reported to increase the probability of personnel injuries. As previously described, the case of China airlines on August 20, 2007 is an excellent example that reveals the negative consequences of inadequate physical access during maintenance tasks and how this situation is critical in generating errors that may lead to serious accidents.

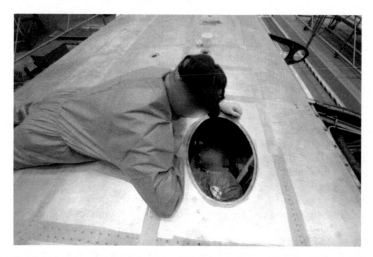

**FIGURE 2.8** Aircraft Confined Space, Fuel Tank. (Source: US Air Force. In public domain. US DoD disclaimer: "The appearance of U.S. Department of Defense (DoD) visual information does not imply or constitute DoD endorsement.")

**Lessons Learned #4**

Aviation maintenance is an extremely demanding task aiming to deliver airworthy aircraft. It is therefore necessary that engineers maintain their situational awareness and follow the appropriate course of action derived from a robust and concrete information process. Frequent knowledge enrichment and training may help to this end.

## REFERENCES

Allen, James. 2013. How good is your hearing? *Aviation MX Human Factors.* 1(4): 7–8.
American Academy of Ophthalmology. 2013. *Introducing Ophthalmology: A Primer for Office Staff.* 3rd ed. San Francisco, Calif.
Australian Transport Safety Bureau. Australian Government. *Aviation Occurrence Investigation AO-2017-099*, released November 14, 2018.
Ballou, Glen (ed.). 2015. *Handbook for Sound Engineers.* 5th ed. Focal Press.
Barron, Randall F. 2003. *Industrial Noise Control and Acoustics.* New York: Marcel Dekker.
Bowling, S. R., S. Kaewkuekool, M. T. Khasawneh, R. Desai, and A. K. Gramopadhye. 2002. *Confined Space Work in Aircraft Maintenance Industry: Scope for Improving Safety and Reducing Errors.* Annual Conference. Proceedings. Institute of Industrial and Systems Engineers. Clemson University.
CAA (Civil Aviation Authority) UK 2002. *An Introduction to Aircraft Maintenance Engineering Human Factors for JAR 66.* Cap 715.
Endsley, M. R. 1995. Measurement of Situation Awareness in Dynamic Systems. *Human Factors: The Journal of Human Factors and Ergonomics Society.* 37: 65–84.
European Parliament and Council. Directive 2003/10/EC of February 6, 2003.
Everest, F. Alton. 2001. *Master Handbook of Acoustics.* 4th ed. McGraw-Hill.
FAA (Federal Aviation Administration). 1998. *Hearing and Noise in Aviation, Medical Facts for Pilots.* AM-400-98/3.

FAA (Federal Aviation Administration). 2014. *Operator's Manual, Human Factors in Aviation Maintenance.*

Goldstein, E. Bruce. 2011. *Cognitive Psychology: Connecting Mind, Research and Everyday Experience.* 3rd ed. Belmont, Calif: Wadsworth, Cengage Learning.

Khurana, A. K. 2007. *Comprehensive Ophthalmology.* 4th ed. New Delhi: New Age International (P) Limited, Publishers.

Lang, G. K. 2000. *Ophthalmology: A Short Textbook.* New York: Thieme Stuttgart.

Martinussen, M., and D. R. Hunter. 2018. *Aviation Psychology and Human Factors.* Boca Raton: CRC Press.

Moore, Brian C. J. 2013. *An Introduction to the Psychology of Hearing.* 6th ed. Leiden: Brill.

Noweir, Madbuli H., and Mohamed A. Zytoon. 2013. Occupational Exposure to Noise and Hearing Thresholds among Civilian Aircraft Maintenance Workers. *International Journal of Industrial Ergonomics.* 43(6): 495–502.

Orasanu, J. and U. Fischer. 1997. Finding Decisions in Natural Environments: The View from the Cockpit. In *Naturalistic Decision Making,* ed. C. Zsambok and G. Klein, 343–357. Hillsdale, NJ: Lawrence Erlbaum Associates.

Smedje, G., M. Lundén, L. Gärtner, H. Lundgren, and T. Lindgren. 2011. Hearing Status among Aircraft Maintenance Personnel in a Commercial Airline Company. *Noise Health.* 13: 364–370.

Solso, Robert L. 2001. *Cognitive Psychology.* 6th ed. Boston: Allyn and Bacon.

Sternberg, Robert J. 1999. *The Nature of Cognition.* A Bradford Book. MIT Press.

Sternberg, Robert J., and Karin Sternberg. 2012. *Cognitive Psychology.* Belmont, Calif: Wadsworth, Cengage Learning.

Tulving, E. 1985. *Elements of Episodic Memory.* Oxford: Oxford University Press.

Wickens, Christopher D., Justin G. Hollands, Simon Banbury, and Raja Parasuraman. 2016. *Engineering Psychology and Human Performance.* London: Routledge.

Wiegmann, Douglas A., and Scot A. Shappell. 2003. *A Human Error Approach to Aviation Accident Analysis: The Human Factors Analysis and Classification System.* Aldershot, UK: Ashgate Publishing.

# 3 Social Psychology

Safe behavior is a prerequisite for non-hazardous performance and a key component of safety culture. However, it is influenced by a number of factors that are researched by *social psychology*, which is a constantly evolving science. According to theory, "people's thoughts, feelings and behaviors are influenced by other people" (Allport 1985), while individuals tend to comply with and conform to group influence (Coultas and van Leeuwen 2015; Cialdini and Goldstein 2004). In addition, researchers suggest that individuals are influenced by social group dynamics to such an extent that they can alter their own behavior.

As shown in Figure 3.1, the field of aircraft maintenance is not self-determined, but rather exists within a system that interacts and modulates the system accordingly. By simplifying the system we can observe two primary lines of influence that affect the workplace of aircraft engineers within the aircraft maintenance system.

The first line of influence is endogenous and regulatory in nature and involves the implementation of norms and procedures regarding the airworthiness of aircraft, the operation and function procedures of maintenance organizations, as well as the professional criteria and eligibility conditions of engineers. This action has been pursued by international, regional and national authorities and has formed *de jure* a highly regulated industry. The consequence of this influence on the daily aircraft maintenance environment is the resulting obligation of maintenance organizations to develop the necessary systemic structures, as well as the obligation of the engineers themselves to behave with a high level of professionalism. The main characteristics of the required behavior are the need for *compliance* with applicable regulations and procedures and the increased *responsibility* and *accountability* that engineers feel and carry, whether they undertake their own tasks or participate as members of maintenance teams.

At the same time, the maintenance working environment is influenced by exogenous factors, which de facto have an impact on the aviation industry. These elements emanate from national, regional and/or global adverse economic and geopolitical changes or other local social conditions that are ultimately reflected in the working environment of engineers, through the demonstration of behaviors ripe with frustration, uncertainty or prejudice.

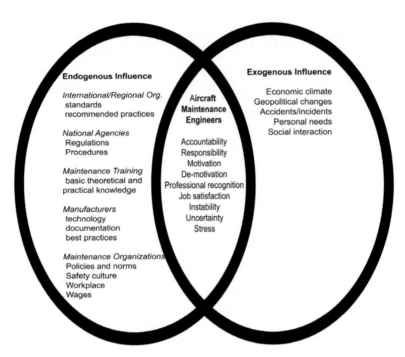

**FIGURE 3.1** Endogenous/Exogenous Influences on Aircraft Engineers. (Source: Created by Yiannakides and Sergiou 2019.)

Hence, before analyzing the key social psychology factors interacting in the aircraft maintenance workplace, it must be acknowledged that this environment is essentially multicultural. Moreover, its consistency and efficiency depends on the behavior of its members, which could under certain circumstances prove to be detrimental to safety performance.

## 3.1 RESPONSIBILITY ATTRIBUTES ON INDIVIDUALS AND GROUPS

The concept of *responsibility*, the way people perceive it and its influence on behavior in the workplace is a basic subject of research in the wider field of psychology. Indeed, *responsibility* is distinguished in various forms depending on its source and content. As a general concept, responsibility can be understood as "a sphere of duty or obligation assigned to a person by the nature of that person's position, function, or work" (Barry 2014), or recently as "the state or the fact of being accountable or to blame for something" (Sheldon et al., 2018).

The modern literature separates *responsibility* and *accountability* by underlining that the latter entails the obligation of the person who assumes *responsibility*, as concerns the practices of reporting, explaining or justifying his actions, whilst a basic separation between the two is that *responsibility* can be shared among individuals, which is not applicable for *accountability*.

A question that is frequently raised from both young engineers and their more experienced colleagues concerns the specific responsibility they assume in their work. The answer to this question can be found in the corresponding Regulation, which defines both the procedures that engineers must follow when carrying out the maintenance of the aircraft, and the procedures applied in their training and licensing. A careful study of the relevant European Regulation demonstrates that *professional* and *legal responsibility* includes accountability: "any person or organisation performing maintenance shall be responsible for the tasks performed" (Section A, Subpart B, M.A 201 Responsibilities), as well as "a person or group of persons shall be nominated with the responsibility to ensure that the organization is always in compliance with this Subpart. Such person(s) shall be ultimately responsible to the accountable manager" (Section A, Subpart F, M.A 606 Personnel Requirements).

It is therefore obvious that the profession of aircraft maintenance engineer involves a high degree of professional and legal responsibility. In addition, by considering the adverse impact of maintenance errors on individuals and the wider society, it is no exaggeration to state that the profession also involves a moral responsibility. However, this aspect necessitates further investigation. Another important issue is that according to the Regulation, the maintenance organization must establish the competency of each individual, which implies that non-certifying technicians also take responsibility in the maintenance process.

Studies in the work environment have shown that the attitude of personnel as regards responsibility attribution depends on whether they work autonomously or as members of teams. Notably, Sheldon et al. suggest that autonomy is beneficial for responsibility and that people are more prepared to take responsibility for their work (2018).

Based on the above, it is evident that the sense of individual responsibility is weakened when distributed between the members of the group, as compared to the case of autonomous work. Therefore, the responsibility-sharing factor must be taken into account when applied in teams, which is now the norm in aircraft maintenance. In addition, the responsibility assumed by engineers as team members requires safety-enhancing behavior that should be characterized by compliance with procedures, cooperation between members and the readiness of each individual to report and justify his actions if needed. Personal integrity and respect are also among the core values that should characterize the behavior of engineers. It should be mentioned that *personality* is another important factor, which determines the level of *perception of responsibility* and shapes the attitude of the individual in the workplace (Robertson and Callinan 1998). To this end, personality traits that are typically shaped at a young age and which are related to various cultural issues should be considered in teamwork.

The most powerful proof of the responsibility assumed by engineers is the authority they have to sign the *Certificate of Release to Service* (CRS). The CRS is signed by a licensed aircraft engineer with the appropriate certifying approvals on completion of maintenance and prior to return to service. This procedure asserts that the aircraft is airworthy to the degree appropriate for safe flight. According to EASA, a CRS shall be issued by appropriately authorized certifying staff on behalf of the organization when it has been verified that all maintenance ordered has been properly carried out by the organization in accordance with the procedures [...] taking into account the availability and use of the maintenance data [...] and that there are no non-compliances which are known to endanger flight safety.

As a minimum, the CRS should contain the basic details and limitations concerning airworthiness or operations, the type of maintenance carried out if any and the identity of the organization and/or person issuing the release to service (EASA 2015, 2014)

In terms of duties, the related responsibilities in aircraft maintenance could be drawn as an aircraft maintenance engineer *base* and as an aircraft maintenance *line* engineer. Duties are based on the privileges attributed by their license and typically involve maintenance required to be performed on aircraft, in order to ensure compliance with the maintenance program and maintain airworthiness.

## 3.2 MOTIVATION INFLUENCE

Work *motivation* has been defined as the internal force that drives employees to realistically work according to their abilities, skills and knowledge and includes all the external factors that bolster this force (Locke and Latham 2002). Work *motivation* has been a major subject in many theories developed over the past 50 years (Maslow [1954] 1970; Alderfer 1973). Although none are supported entirely from the results and more research is needed, they all converge to the conclusion that motivation influences organizational behavior and improves performance (Aamodt 2010; Kanfer and Chen 2016).

In Maslow's theory, it is suggested that humans are driven by two different sets of motivational forces that must ensure survival and self-actualization needs. Maslow presented a five-level pyramid that contains the needs that people have in a bottom-up hierarchical sequence. He theorized that physiological/biological needs (food, water etc.) listed at the bottom should be satisfied first, followed by the need to feel secure. Then we find the need to achieve social integration by enjoying love and friendship, followed by the satisfaction of ego via the acquisition of recognition and success and finally the level where we reach self-actualization, which means that we have accomplished all the challenges of life.

Alderfer, on the other hand, in his ERG theory followed a three-level approach. He noted that people have to fulfill the needs derived from their *existence* such as salary, job and security. The next level concerns *relatedness*, which encompasses social and external esteem, including friendships, relationships, common goals and recognition and finally we have *growth* that fulfills internal esteem and self-actualization. The biggest difference between the two theories is that Alderfer suggests that the levels are not necessarily sequenced hierarchically, but, rather, people have the ability to skip needs at various levels.

Motivation can be intrinsic, which means that individuals need to perform well, especially those with a high self-esteem, enjoying or challenged by their work (Deci and Ryan 2014). It can also be extrinsic, which refers to the incentives provided by the employers such as pay, promotion, career opportunities etc. In terms of extrinsic motivation, studies suggest that in order to achieve job satisfaction and drive employees to behave in a way that enhances performance, incentives should be provided to such an extent that they maintain the principle of *equity theory* (Adams 1965). According to this theory, workers feel satisfied only when their work input such as studies, skills and knowledge equals their corresponding output, such as salary, job benefits and responsibilities. Therefore, based on the equity theory, management and especially supervisors could provide the necessary incentives to minimize the input–output gap.

### 3.2.1 MOTIVATION IN AIRCRAFT MAINTENANCE

In aircraft maintenance, motivation success depends on the behavior of management and supervisors, which should aim to satisfy engineers' intrinsic needs by enhancing confidence, facilitating task processing and attributing responsibility according to skills and knowledge. Supervisors must behave with transparency, especially in task allocation and shift planning, in order to establish better communication and understanding among maintenance teams. At the level of management, extrinsic motivation should be provided in terms of rewards or promotion opportunities, in order to create the conditions for engineers to achieve their personal and professional ambitions. Finally, some characteristics of the aircraft engineer profession act as significant work motivators and consequently regulate the behavior of engineers. These include the derivative responsibility and accountability, the completion of a complex maintenance task and the release of aircraft for safe flight, the social perception of the profession and the severity of the consequences of errors.

On the other hand, due to the nature of the industry, there are numerous factors that could be perceived as negative motivators that threaten *existence* and *growth*. For example, the fear of losing work due to external factors that threaten the industry, or even more internal issues such as the working environment or relationships between colleagues, can be a negative incentive and cause a change in behavior. In this regard, particular attention should be paid to provide the appropriate treatment. It should be noted that negative motivation can create workplace conflict, capable of threatening safety, since it affects decision-making and task performance through disorientation, disturbance and uncertainty to individual or team tasks.

## 3.3 PEER PRESSURE IMPACT

Peer pressure is the pressure individuals receive from colleagues or supervisors and the tendency to modify their behavior, as an escape strategy to avoid rejection or to gain social approval. Studies have shown that individual behavior and performance depend on those of other members of the team which create positive or negative motivation (Kandel and Lazear 1992; Cornelissen et al. 2013). Peer pressure can be generated by colleagues' interaction during the accomplishment of a common task, or just by observing each other without interacting at all. Although research in peer pressure focuses on specific occupations, the repercussions on work performance and their extent is still unclear.

### 3.3.1 INCIDENT/ACCIDENT EXAMPLE 8: THE CASE OF AMERICAN AIRLINES, MAY 25, 1979

In aviation, tragedies can occur when organizational unsafe practices and peer pressure are not identified and eliminated in a timely manner. On a corporate level, there have been significant events leading to deviations in safety standards, in favor of excessive financial gain. The crash of American Airlines Flight 191 in Chicago was precipitated by events and procedures that were triggered and implemented by the airline's upper-level management.

According to the accident investigation report (NTSB 1979), 258 people on board the aircraft and two people on the ground were killed when a McDonnell-Douglas DC-10 crashed, shortly after take-off, because the left engine detached from the aircraft. The causal factor for the accident was the difficulty to detect a crack at the engines pylon that progressively propagated and eventually caused the detachment. Contributing to the causal factors was the management's endorsement to use a forklift that was a non-approved procedure, in order to change engines on its fleet, with the aim to expedite maintenance procedures and keep the aircraft in the air, thus generating more profit. Another contributing factor was the engineers' conformity to peer pressure effect and obvious bad organizational norms.

## 3.4 THE INFLUENCE OF CULTURES ON SAFETY

Safety culture has been described as "the engine that continues to drive organizations towards the goal of maximum attainable safety of current commercial pressures or who is occupying the top management posts" (Reason and Hobbs 2003). It constitutes a common set of values and attitudes that promote safety, which are shared by every member on all levels of an organization, but that, at the same time, requires commitment, awareness and willingness to learn from the past.

Successful adoption of a *safety culture* presupposes an understanding of the impact of main cultural factors on employees' behavior, which interact in the field of aircraft maintenance. The analysis of the factors lies in the field of *cross-cultural psychology*, which examines the attitudes, behaviors and moral values of individuals (Ho and Wu 2001). Ongoing research on the interaction between individuals of different ethnic origins, as well as their interface with procedures, machines and regulations, supports the existence of significant differences on the individual level, as concerns the understanding of important aspects of safety, which consequently challenge the working environment (Merritt and Maurino 2004).

The three main cultures that continue to be the subject of study, since they cause divergent behaviors, are *organizational, national* and *professional* cultures. Although there has been considerable convergence on a global level, standardization has not been achieved either in education or in the structure and operation of aircraft maintenance organizations (Helmreich and Merritt 1998).

### 3.4.1 ORGANIZATIONAL CULTURE

*Organizational culture* can also be found in literature as "corporate culture" or "corporate personality" (Dempsey and Gessel 1997). It concerns the social and psychological environment where employees behave and act. It can be measurable since, among others, it is mirrored through the stakeholders', clients' and employees' perception of the organization. It includes the values and norms, policies and procedures, determines behavior and practices and sets safety goals and tolerance levels. The effectiveness of organizational culture is based on management's commitment to promote and monitor safety (Booth 1996). Furthermore, it has the potential to affect individual-to-individual and personnel-to-management interaction, as well as information sharing and occurrence reporting, teamwork performance and error identification and mitigation.

### 3.4.2 PROFESSIONAL CULTURE

*Professional culture* determines the attitude and capabilities of a specific professional group. It begins to form during studies and takes its shape at employment, while its aspects are visible beyond the workplace in the wider social environment. It would be extremely interesting if surveys examined the common features in the professional behavior of aircraft engineers, in a similar manner that existing studies assess the professional behavior of pilots, to determine the similarities they display in their assessment and risk response. Such studies would definitely contribute to the promotion of safety. *Professional culture* evolves through on-the-job training and the accumulation of experience and is consistent with the professional path and the organizational culture. In the context of aircraft maintenance, the culture of engineers is affected by the introduction of safety-related sub-cultures such as *just culture* that is a prerequisite for an effective occurrences reporting system.

### 3.4.3 NATIONAL CULTURE

The airline industry is nowadays characterized by cultural diversity. Similarly, aircraft maintenance has become a multicultural workplace for engineers of different nationalities and ethnicities. This environment has benefited the industry with valuable experience and knowledge sharing, but has also increased the difficulty in communication and cooperation, due to the absence of a common language and code of understanding.

Due to the diversity of the culture, two behavioral factors have been identified that, if not taken into account, may affect the safe handling of maintenance activities. The first factor concerns the differences in understanding the power of leadership in the workplace, as well as the fact that engineers from specific ethnicities never question the instructions of supervisors or of more experienced colleagues, even when they realize that they are incorrect or risky. The second factor concerns teamwork, since it is observed that engineers from a particular ethnic group prefer individual work, thus challenging the quality and the results of teamwork. As a result, cultural diversity creates differences in behavior towards teamwork cooperation, adaptation to changes and compliance to procedures and regulations that may challenge safe performance (Amaram 2007).

## 3.5 TEAMWORK CHARACTERISTICS AND BEHAVIORS

It is truly remarkable that *teamwork* has been a field of research for many scientists in an effort to allocate the most appropriate definition. Furthermore, studies have assisted in defining the behaviors members need to develop, in order to enhance *teamwork* effectiveness. The majority of definitions suggest that in order to have a team, two or more persons should be assigned to it and all members need to cooperate to achieve a shared goal.

Salas et al., for example, following a meta-analysis approach delineate the critical considerations that should be taken into account by the supervisors when they manage groups (2014). They suggest nine considerations, which as a memorable heuristic all begin with "C": *cooperation, conflict, coordination, communication, coaching, cognition, composition, context* and *culture*. Other important factors such as *commitment* and *interdependence* (Francis and Young 1979) and trust (Mathieu et al. 2008) are also identified in studies. Another factor that should be considered carefully is that the more stable a team is, in terms of the duration that the members have worked together, the better its overall performance becomes. However, it should be noted that individuals may tend to consume less effort than when working autonomously, a condition known as *social loafing* (Karau and Williams 1993), which should also be considered by assigning members to teams and monitoring their overall contribution.

The backbone of a safe and reliable air transportation system are the aircraft maintenance engineers (Kraus and Gramopadhye 2001) whose performance and effectiveness nowadays relies mainly on teamwork. Consequently any deficiencies in *communication* or *conflicts* that pop up in *decision-making* must be identified and solved immediately. Knowledge and experience sharing and *assertive behavior*, through constructive criticism of others and ourselves, are also key components for *undisrupted communication* and *decision-making*. Last but not least, fair workload allocation and provision of adequate resources, tools and operating procedures facilitate team members' efforts and limit the likelihood of *fatigue* symptoms, *stress* and task *overload*.

Finally, based on empirical evidence, team composition is of paramount importance in aircraft maintenance, since in many cases a task that a team assumes is not completed in one shift but is transferred and taken over by the next shift consisting of other members. This requires that supervisors build maintenance teams in such a way that the accumulated experience and capacity are distributed equally. At the same time, the transfer of work, which implies a sharing of responsibility, requires *trust* not only between members of the same group but also between members of different groups. This particular feature of aircraft maintenance teams is found in only a few professions where *trust* and *accountability* is achieved through detailed briefings and de-briefings between groups and individuals, which must be accompanied by the delivery of the signed task sheets.

## 3.6 MANAGEMENT, SUPERVISION AND TEAM LEADERSHIP

According to Thompson and Strickland (1996), "Managers must combine good strategy-making with good strategy execution for company performance to approach maximum potential by converting strategic vision, mission into measurable objectives and performance targets". Strategy implementation includes, inter alia, budgeting, employment requirements, resources procurement, regulations compliance and the creation of an organizational culture with the appropriate motivations, best practices and norms. This role is assigned to managers who often assume covert work and roles and act at the top level of organizational structures.

In contrast, supervisors act on the operational level and are very visible and approachable to other employees. Due to their personal knowledge, skills and experience, they are responsible for setting task goals and deadlines, in order to implement the management's strategy. To this end, they monitor productivity flow by composing and supervising the work teams and provide feedback to the management. They are also the physical link of communication between employees and management by transferring employees' requests and complaints (upward communication) or the management's instructions, goals and benefits (downward communication). Undoubtedly, managers and supervisors exercise leadership and their prevalent behavior shapes the social framework within which workers are required to coexist and work. Figure 3.2, depicts some of the basic characteristics of effective leadership.

In the 1950s, a strong research activity was observed in the field of leadership, but it was some decades later that emphasis was put on the particular characteristics of the leader's behavior and on their corresponding influence on working groups (Lord et al. 2017). Although the results of many researchers are conflicting, five personality traits that drive leadership behavior are dominant, known in the field of *personality psychology* as the "Five Factor Model" (Judge et al. 2002; Northouse 2016). First, leaders who are characterized by *neuroticism* and *extraversion* are found to be attentive to the team's needs, whilst they are more social, assertive and energetic. Second, the individuals who display *openness* are more capable of finding creative solutions, while *agreeableness* characterizes leaders who are trusting, compliant, sympathetic and cooperative. Finally, the leaders with *conscientiousness* assume responsibility and accountability and achieve targets with diligence.

In the field of aircraft maintenance, the behavior of managers should be consistent with their stated policy of promoting safety. It is therefore their responsibility to comply with regulations and to create a positive organizational culture. Moreover, they should aim to provide incentives and to treat all workers fairly. Respectively, supervisors have to exercise their leadership in an exemplary manner, encourage occurrence reporting and prevent the establishment of negative norms, keeping in mind that their behavior determines the successful or failed performance of work teams. The legacy of James Reason prevails today, since in the event of an accident/incident, the actions undertaken by management, which actually created the latent unsafe conditions leading to active failure, are no longer ignored or hidden.

### Lessons Learned #5

A plethora of social factors in the working environment of aircraft maintenance have the potential at all times to nurture behaviors that can be detrimental to safety. Assume a hypothetical working environment where

**FIGURE 3.2** Effective Leadership. (Source: AdobeStock. ©gustavofrazao – stock.adobe. com. Under Standard License AdobeStock_104382742.)

engineers are responsible and accountable and apply safe and best practices that are reinforced and supported by the management. In this environment engineers perform better and promote safety. On the other hand, if the management does not provide motivation and the necessary resources, while supervisors permit shortcuts and "out of the book" practices, then the outcome will likely be breaches in safety.

## REFERENCES

Aamodt, Michael G. 2010. *Industrial/Organizational Psychology: An Applied Approach.* Belmont, Calif: Wadsworth Cengage Learning.

Adams, J. S. 1965. Inequity in Social Exchange. In *Advances in Experimental Social Psychology*, ed. L. Berkowitz, 267–299. New York: Academic Press.

Alderfer, Clayton P. 1973. *Existence, Relatedness, and Growth: Human Needs in Organizational Settings.* New York: Free Press.

Allport, A. 1985. The Historical Background of Social Psychology. In *Handbook of Social Psychology*, ed. G. Lindzey and E. Aronson, 1–46. New York: Random House.

Amaram, Donatus I. 2007. Cultural Diversity: Implications For Workplace Management. *Journal of Diversity Management.* 2(4): 1–6. https://doi.org/10.19030/jdm.v2i4.5017.

Barry, Vincent. 2014. *Moral Issues in Business.* Belmont, Calif: Cengage Learning.

Booth, R. T. 1996. The Promotion and Measurement of a Positive Safety Culture. In *Human Factors in Nuclear Safety*, ed. N. A. Stanton, 313–332. London: Taylor & Francis.

Cialdini, R. B., and N. J. Goldstein. 2004. Social Influence: Conformity and Compliance. *Annual Review of Psychology.* 55: 591–621.

Cornelissen, Thomas, Christian Dustmann, and Uta Schönberg. 2013. *Peer Effects in the Workplace.* Munich: CESifo.

Coultas J., and E. van Leeuwen. (2015) Conformity: Definitions, Types, and Evolutionary Grounding. In *Evolutionary Perspectives on Social Psychology. Evolutionary Psychology*, ed. V. Zeigler-Hill, L. Welling, and T. Shackelford. Cham: Springer.

Deci, Edward L., and Richard M. Ryan. 2014. *Intrinsic Motivation and Self-Determination in Human Behavior*. New York: Springer Science+Business Media.

Dempsey, Paul Stephen, and Lawrence E. Gesell. 1997. *Air Transportation: Foundations for the 21st Century*. Chandler, Ariz.: Coast Aire.

EASA (European Aviation Safety Agency). 2014. Part M, Subpart H, Certificate of Release to Service — CRS, M.A.801 Aircraft certificate of release to service and EASA Part 145.A.50 Certification of maintenance [EC, 2014].

EASA (European Aviation Safety Agency). 2015. *EASA Policy on Certificates of Release to Service for Aircraft Maintenance and Associated Responsibilities of Maintenance Organisations and CAMOs*.

Francis, D., and D. Young. 1979. *Improving Work Groups*. San Diego, Calif.: University Associates.

Helmreich, R., and A. Merritt. 1998. *Culture at Work in Aviation and Medicine: National, Organizational and Professional Influences*. Brookfield, VT: Ashgate.

Ho, D. Y. F., and M. Wu. 2001. Introduction to Cross-Cultural Psychology. In *Cross-Cultural Topics in Psychology*, ed. Leonore Loeb and Uwe P. Gielen, 3–13. Westport, CT: Praeger.

Judge, Timothy A., Joyce E. Bono, Remus Ilies, and Megan W. Gerhardt. 2002. Personality and Leadership: A Qualitative and Quantitative Review. *Journal of Applied Psychology*. 87(4): 765–780.

Kandel, Eugene, and Edward P. Lazear. 1992. Peer Pressure and Partnerships. *Journal of Political Economy*. 100(4): 801–817.

Kanfer, Ruth, and Gilad Chen. 2016. Motivation in Organizational Behavior: History, Advances and Prospects. *Organizational Behavior and Human Decision Processes*. 136: 6–19.

Karau, S., and K. Williams. 1993. Social Loafing: A Meta-Analytic Review and Theoretical Integration. *Journal of Personality and Social Psychology*. 65(4): 681–706.

Kraus, David C., and Anand K. Gramopadhye. 2001. Effect of Team Training on Aircraft Maintenance Technicians: Computer-Based Training versus Instructor-Based Training. *International Journal of Industrial Ergonomics*. 27(3): 141–157.

Locke, E. A., and G. P. Latham. 2002. Building a Practically Useful Theory of Goal Setting and Task Motivation: A 35-Year Odyssey. *The American Psychologist*. 57(9): 705–717.

Lord, Robert G., David V. Day, Stephen J. Zaccaro, Bruce J. Avolio, and Alice H. Eagly. 2017. Leadership in Applied Psychology: Three Waves of Theory and Research. *Journal of Applied Psychology*. 102(3): 434–451.

Maslow, A. H. [1954] 1970. *Motivation and Personality*. New York: Harper & Row.

Mathieu, John, M. Travis Maynard, Tammy Rapp, and Lucy Gilson. 2008. Team Effectiveness 1997–2007: A Review of Recent Advancements and a Glimpse Into the Future. *Journal of Management*. 34(3): 410–476.

Merritt, A., and D. Maurino. 2004. Cross-Cultural Factors in Aviation Safety. *Advances in Human Performance and Cognitive Engineering Research*. 4: 147–182.

Northouse, Peter Guy. 2016. *Leadership: Theory and Practice*. Thousand Oaks: Sage.

NTSB (National Transportation Safety Board). Report No. AAR-79/17, released December 21, 1979. Retrieved from Aviation Safety Network. Flight Safety Foundation.

Reason, James, and Alan Hobbs. 2003. *Managing Maintenance Error: A Practical Guide*. Aldershot, UK: Ashgate.

Robertson, Ivan, and Callinan, Militza. 1998. Personality and Work Behaviour. *European Journal of Work and Organizational Psychology*. 7(3): 321–340. doi:10.1080/135943298398736.

Salas, E., M., L. Shuffler, A. L. Thayer, W. L. Bedwell, and E. H. Lazzara. 2014. *Understanding and Improving Teamwork in Organizations: A Scientifically Based Practical Guide*. Wiley Periodicals. doi:10.1002/hrm.21628.

Sheldon K. M., T. Gordeeva, D. Leontiev, M. F. Lynch, E. Osin, E. Rassakazova, and L. Dementily. 2018. Freedom and Responsibility Go Together: Personality, Experimental, and Cultural Demonstrations. *Journal of Research in Personality*. 73: 63–74.

Thompson, Arthur A., and A. J. Strickland. 1996. *Strategic Management: Concepts and Cases*. Boston, Mass: Mcgraw-Hill.

# 4 Factors Affecting Performance

This chapter examines individual factors that affect employee performance and thus have the potential to result in unsafe actions. The factors fall into the field of psychophysiology since they can have physical or mental effects. In order to remain focused on the purposes of this book, emphasis is given to the mental effects and the impact on the process of information processing and, in particular, the decision-making process, both of which are discussed in Chapter 2. The prevailing factors of *time pressure, workload* and *shift work* and the consequent symptoms of *stress* and *fatigue* are examined within the working environment of aircraft maintenance, since they continue to be the dominant human factors leading to unsafe acts, thus contributing to faulty actions and errors from engineers. Although these factors are mentioned separately, so as to be more apprehensible, it should be noted that they often arise from each other and therefore coexist.

## 4.1 TIME PRESSURE

As previously highlighted, aircraft maintenance has become time constrained with an unquestionable and ongoing demand for aircraft utilization and on-time departures. Consequently, *time pressure* has historically constituted one of the most severe stressors and is the latent failure contributing to aircraft incidents and accidents. Moreover, aircraft delays occurring due to specific factors of the aviation system, such as the impact of meteorological conditions and congested airspace, further reduce the available time allocated for aircraft maintenance between scheduled take-offs and thus escalate *time pressure.*

*Time pressure* is the psychological reaction of individuals when they face time constraints when performing a task (Ordóñez et al. 2015). This reaction is mainly imposed by supervisors and managers, but it could also be self-imposed, since it is experienced by engineers on some occasions, even when no strict deadlines or time constraints have been set. Prior research supports the notion that *time pressure* affects the decisions taken by individuals (MacGregor 1993) and impairs performance because it places constraints on the capacity for thought and action by influencing the cognitive effort for information search and processing to a certain extent (Moore and Tenney 2012).

Moreover, *time pressure* seems to be more intense when individuals are dealing with multiple tasks. However, the way individuals choose to perform multiple tasks under pressure is yet not clear. Some researchers suggest that individuals dealing with multiple tasks tend to complete the ones imposed by time constraints to the exclusion of others (Leroy 2009), meaning that error opportunities are created due to task interruption. On the other hand, some studies argue that time pressure may trigger the phenomenon of cognitive lockup, thus rendering individuals reluctant to switch to another task, even if that task has a higher priority (Schreuder and Mioch 2011). Based on an alternative approach, researchers suggest that the way individuals perceive *time pressure* and the derivative impact on performance varies from individual to individual, according to *personality* characteristics and *cultural* differences. Perhaps this explains why maintenance organizations operating under constant *time pressure* do not experience many incidents compared to other organizations operating at normal rates.

In view of further research, existing studies reveal that 25 percent of unsafe acts committed in aircraft maintenance involve *time pressure* (Suzuki et al. 2008). The effects of time constraints also influence supervisors, who find it difficult to make the right decisions, probably because they become less creative, which is an imperative personality trait included in the Five Factor Model (see Chapter 3). Although there are studies that suggest that *time pressure* can progressively enhance the process of *information processing*, when we take into account the complex and multitasked environment of aircraft maintenance, it is evident that the influence of *time pressure* on risky behaviors must be acknowledged, whilst further efforts for continued training should take place, in order to mitigate the negative effects (Kobbeltvedt et al. 2005).

## 4.2 WORKLOAD

Existing research on *workload* has resulted in the development of various methodologies, aiming to evaluate the ability of operators to perform the tasks assigned. At the same time, laboratory and field experiments have demonstrated the need to include a large number of factors in the evaluation process, which are often based on the researcher's observations and on the subjective view of the operator. The nature of work (difficulty, complexity, single or multitask), the skills and competence of the operator, the characteristics of the working environment, as well as the time available to complete the work, are among the basic variables that determine the *physical* and *mental workload* the operator has to undertake, in order to complete the expected work effectively. Whenever the variables change, they produce a new workload that impacts performance and alters the expected result.

Particular emphasis has been placed on *mental workload* (Moray and Nato Special Program Panel on Human Factors 1977; Wickens and Hollands 2016b) and in particular on *high workload* (*overload*), a situation in which the human brain has limited capacity to respond as usual, resulting in the symptoms of stress and fatigue (Smith and Smith 2017), which lead to a slow task performance, risky behaviors and errors.

*Mental workload* has also been found to impact individual performance through its effect on *arousal* (Charlton 2002; Son et al. 2011), which can be described as the level of an individual's alertness and situational awareness. From the Yerkes–Dodson law and work further improved by other researchers (Yerkes and Dodson 1908; Hebb 1949; Diamond et al. 2007), as well as from the empirical inverted U-curve (the Hebbian version) depicted in Figure 4.1, it is clear that there is a point of optimum arousal, under which individuals working on *complex and difficult tasks* can achieve top performance. In conditions where workload levels are reduced, individuals may feel bored and not able to perform as expected. On the other hand, whenever individuals perform under high workload and arousal they experience an inverse effect and subsequent detriment in performance. At both edges of the inverted U-curve, there are opportunities for error, since mental information processing either subfunctions or lacks capacity.

### 4.2.1 SCHEDULED AND UNSCHEDULED MAINTENANCE

*Workload* in aircraft maintenance can be both physically and mentally demanding, deriving from scheduled or unscheduled maintenance tasks, whilst it can also demand more mental effort in comparison to physical abilities. *Scheduled maintenance* is projected in aircraft maintenance manuals and is required periodically. To this end, inspections or repairs are performed within the limits of specific flight cycles, hours flown and/or time expired (A, B, C, or D check). Although aircraft manuals normally refer to the required *workload* in terms of resources and man-hours, maintenance organizations are only able to estimate *workload* when adequate experience has been gained through the repetition of tasks. However, scheduled maintenance tasks often become overloaded due to time constraints, manpower shortages, lack of resources and the emergence of unexpected tasks.

On the other hand, *unscheduled maintenance* involves work in response to service bulletins from manufacturers, safety directives from regulatory and oversight authorities, feedback from flight crews and findings from routine aircraft inspections (Wells 1998). Empirically, evidence shows that the *workload* involved for nonroutine maintenance cannot be scheduled with accuracy and demands as much as 50

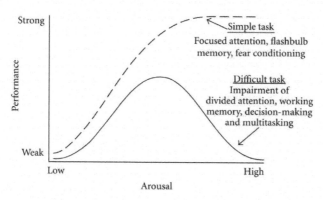

**FIGURE 4.1** Yerkes-Dodson Curve of Arousal (1908) – The Hebbian Version. In public domain.

percent of the planned effort. Therefore, *unscheduled maintenance* often results in *overload,* since it is accompanied with time constraints and the need for more manpower and resources, amid uncertainty, lack of information and knowledge gaps.

## 4.3 SHIFT WORK

*Shift work* has become a common working practice in aircraft maintenance, due to the ongoing demand for 24/7 aircraft airworthiness, but also in order to address additional workload and lack of time from unscheduled maintenance. In addition, engineers are frequently obliged to receive work in progress from colleagues, since the complexity of the systems, the lack of tools and materials and aircraft aging, often prevent completion of maintenance within the working time of a single shift.

Research in different domains where shift work is prevalent reveals that *shift work* can cause physical and mental disorders like obesity and stress (Luckhaupt et al. 2014) and thus impact health, social life, work performance and effectiveness (Saksvik et al. 2011). Particularly in the field of aircraft maintenance, it is argued that shift workers who undertake their tasks in the very early hours of the morning are more likely to commit errors than those following normal schedules (Hobbs et al. 2010). Furthermore, rotating *shift work* in aircraft maintenance that requires engineers to constantly switch from an eight-hour day to an eight-hour night shift is reported to be associated with *sleep deprivation* and *circadian rhythm* disruption, which causes sleepiness and cumulative and circadian *fatigue,* as presented later in this chapter.

*Sleep deprivation* can result from continuous night shifts or work for prolonged periods and has been reported to impair cognitive functioning, thus leading to significant personal and economic consequences, associated with human error and accidents caused by performance deficits (Van Dongen and Dinges 2005; Van Dongen 2006; Haidarimoghadam et al. 2017). In order to understand the magnitude of the problem of *sleep deprivation* in aircraft maintenance, we can merely refer to an FAA field study in aircraft maintenance organizations which argued that the average amount of sleep for engineers was about five hours (Bedell-Avers et al. 2011). In particular, the finding of the study revealed that engineers, in the absence of regulation, often chose to work even beyond their regular shift time, in an effort to secure higher pay, thus overlooking the effects of sleep deprivation and fatigue in performance, which is particularly worrisome.

Although research continues, it turns out that shift workers experience longer sleep disruptions and increased difficulties in replenishing the accumulated *sleep deprivation.* The duration of sleep depends on our internal body rhythm but is often affected by outside stimuli. Figure 4.2 suggests that sleep progressively passes through four stages: stages 1, 2 and 3, which involve the more restorative stage, and Rapid Eye Movement or REM sleep, most commonly known as the dreaming stage. In a complete sleep cycle of an average of 90 to 110 minutes, each stage lasts between five to 15 minutes. The sleep/awake system is often mistakenly considered as a credit/debit system, where an eight-hour sleep (credit) allows for a 16-hour active period (debit). In simple terms, it implies that one hour of sleep can credit the body with two hours of being awake. However, 10 to 12 hours of sleep, following

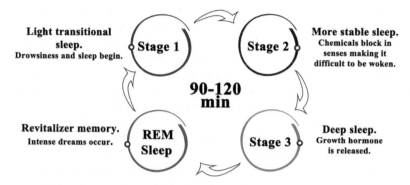

**FIGURE 4.2** Sleep Stages. (Source: AdobeStock. ©Dmitry – stock.adobe.com. Under Standard License AdobeStock_178424203.)

a long period of strenuous activity, will only credit the body with eight hours and the individual will feel sleepy again after 16 hours. Thus, sleep cannot be stored in anticipation of a long period of work.

Researchers believe that the impact of shift work on sleep disruption is associated with the human *circadian rhythm* and the derivative effects may be responsible for health disorders, reduced alertness, poor productivity and fatigue-related accidents (Akerstedt 1998; Drake and Wright 2011). *Circadian rhythm* is the human 24-hour internal biological clock and although it is endogenous, it adapts to the local environment through external factors such as daylight and darkness. Every disturbance of circadian rhythm and, in particular, any successive alteration of the sleep–wake schedule as concerns the external environment, affects a large part of human mood and behavior, as well as mental and physical functions.

Researchers have found that body temperature regulates the body's *circadian rhythm* and influences human performance. Consequently, better performance is achieved when body temperature is high and near the circadian cycle and worse when body temperature is low and nearing the nadir of the circadian cycle (Wright et al. 2002). The lowest performance occurs at around 0500 hours in the morning at a time when the temperature of the body is at its lowest and increases progressively until 1800 hours in the afternoon when it drops again.

However, it should be noted that empirical research results suggest that the level of the circadian cycle desynchronization is also influenced by a number of individual factors, such as age and sex, further to the effect of light exposure that fundamentally depends on the geographic location of workplace (Adan et al. 2012). Consequently, night shift work appears to have a greater impact on the circadian cycle of women and the elderly, thus making them more vulnerable to health disorders and performance decrements.

## 4.4 STRESS

Although a vast number of definitions of *stress* are available depending on the area of interest, the converging view is that both positive as well as negative events can generate stress that emerges as a normal reaction of humans to any physical or mental demands and changes. When the subject under research is *stress* in the workplace it is very typical to divide it into two main categories: *domestic stress* that results from conditions at home; and *occupational* or *job stress* that is created by the general work environment. In order to remain aligned with the scope of this book, and since *stress* generating factors (*stressors*) are widely examined in its chapters, particular emphasis is placed on work-related stress and its potential effect on the behavior and performance of workers.

*Occupational stress* results from the interaction of employees with their working environment and is widely perceived as a reaction that exceeds human adaptive capacities and consequently threatens well-being. Individual differences such as personal resources and work demand have a significant effect on the level of stress and consequently on health outcomes (Mark and Smith 2008). Research results also argue that human response to *stress* differs from individual to individual and can become a predictor of performance during critical incidents (Vine et al. 2015). Despite the fact that *stress* elimination may not be feasible, researchers suggest that it can be managed if individuals identify the source and the magnitude of the stressful situation and take a rational look at the stressors (Homan 2002).

Following a simplistic approach, it could be suggested that stress in aircraft maintenance stems from three main factors: the "engineer"; the "task"; and the "environment" in which the task is performed. The "engineer" factor includes the personality traits and culture of the person involved, as well as the physical and mental resources provided, mainly the skills, the knowledge and the competence to apply them to the specific task. The "task" factor can include the nature of the work, especially whether it is repetitive or not, simple or complex and the level of time constraints and workload involved. Finally, the "environment" factor can include the social interaction of the person with other engineers and supervisors and the physical conditions of the workplace, such as temperature, lighting and noise, which will be further analyzed in Chapter 5.

Despite the prevailing impression that stress only has negative effects on performance, researchers suggest that stress can, to some extent and depending on the individual and the nature of work, increase efficiency. This occurs up to an optimum level of stress, after which the effects of stress generate a negative effect, in line with the previously mentioned Yerkes and Dodson inverted U-curve. It is therefore evident that every engineer should be able to identify and acknowledge his/her personal optimum point of stress and thus develop stress coping strategies, whenever the existing stressors create conditions that have the potential to affect efficient performance or lead to precarious behavior and actions.

Excessive *stress* can be identifiable through the behavior of individuals, since it is characterized by uncertainty and tendency to prolong certain tasks, anxiety, confusion as well as denial and aggression. Behavior may become impulsive and restless, while mood swings may lead to apathy, fatigue and low self-esteem. However, the complete elimination of excessive stress is practically impossible, given that in addition to the factors that create it, a fundamental determinant is the particular characteristics of the individual.

## 4.5 FATIGUE

The term *fatigue* can refer to physical weariness, emotional exhaustion and the degradation of skill resulting from a mentally demanding task over an extended period without an adequate rest (Hobbs 2008). To this end, fatigue can involve physiological and mental impairment, causing a decline in performance. *Fatigue* should not be confused with sleepiness, since workers, even when they are not sleepy, may experience *fatigue* symptoms that are likely to affect their performance (Phillips 2014). However, there is sufficient evidence suggesting that cumulative *sleep deprivation* can accelerate the impact of *mental fatigue*.

When assessing the safety implications of *mental fatigue* in aircraft maintenance, research results suggest that people in this situation are unable to maintain the same quality in their cognitive processing, including a reduced ability to assess risk or appreciate consequences of actions and reduced *situational awareness*. Furthermore, as the levels of mental fatigue increase, workers display an aversion to task performance (Lorist et al. 2000), through poor attention, decreased memory capabilities and impairment of judgment. Indeed, an FAA survey conducted in 2006 on international human factors programs in maintenance organizations showed that more than 80 percent of the respondents believed *fatigue* had been an issue (Bedell-Avers 2011).

Since *fatigue* has been reported as one of the main contributing factors for unsafe acts, it is now recognized as a core safety issue in aviation. To this end, the ICAO and other safety authorities promote *Fatigue Risk Management System* (FRMS) programs as part of the wider spectrum of safety management systems (SMS) that will be further analyzed in Chapter 7. FRMS involves the safety policies and practices of maintenance organizations and requires the operationalization of a safety occurrences reporting system that employees will use with trust, in order to become a practical decision-making tool which can assist them in planning and allocating tasks in an orderly and proactive manner, so as to avoid detrimental levels of *fatigue* (Hobbs et al. 2011). Ideal FRMS utilization can also contribute to the reduction of work-related stress, high *workload* and *time pressure* by predicting fatigue risk, thus maximizing personnel alertness and improving safety performance.

Transport Canada has introduced a valuable fatigue risk management toolbox that suggests a number of tools that organizations and individuals may exploit in order to educate themselves on the nature and impact of fatigue and eventually assess fatigue risk. In addition, the FAA suggests three types of fatigue that are mainly associated with prolonged labor and sleep deprivation, aiming to increase the awareness and education of the specific symptoms, in order for them to be identifiable (FAA 2012). These are *transient fatigue*, that refers to acute fatigue, *cumulative fatigue*, that is brought on by repeated mild sleep restriction or extended hours of being awake across a series of days and *circadian fatigue*, that involves reduced performance during night-time hours.

## 4.6 COPING WITH STRESSORS

Studies have shown that physical activity can combat low energy and feelings of fatigue (Puetz 2007), while physical exertion can help individuals control excessive stress levels and improve physiological and brain functions (Chaouloff et al. 2013). Aging and obesity have continuously been reported as latent medical factors that impose risks to personal health and aircraft safety. Thus, it is important for engineers to maintain their health and fitness at optimal levels. In order to achieve that, they have to adopt regular physical exercising and hygiene habits, and also abstain from adverse use of drugs and alcohol. They should be fully aware of the negative impacts of drug and alcohol use on their health and of the resulting safety risks. Moreover, they should be mindful of the right of authorities and maintenance organizations to conduct regular and post-accident drug and alcohol tests on all workers who perform sensitive duties, without prior notification, in accordance with FAA and EASA Regulations. Subsequently, a license could be revoked, suspended or limited, in the event that an engineer has been found to be under the adverse influence of alcohol or drugs when carrying out maintenance or issuing a certificate of release to service. As a general rule, engineers should not work for at least eight hours after drinking even small quantities of alcohol, whilst they should increase this time if more quantities have been consumed. In addition, they must refrain from the use of any illegal drugs. Even if the drugs are legal and intended for therapeutic purposes, they should consult their doctor on the associated symptoms, so as to acknowledge the possible impact on their physical and mental condition.

The case of the fatal accident of Emery Worldwide Airlines, Flight 17, is an example of the application of the regulations. A McDonnell Douglas DC-8–71F crashed on February 16, 2000 due to pitch control loss following an elevator malfunction, shortly after taking off from Sacramento Mather Airport. According to the investigation report (NTSB 2003), although not required at the time, voluntary drug tests were executed between one and six days following the accident that were positive for two employees who were subsequently relieved of their duties, *despite the fact that it was determined that their performance was not a contributing factor to the accident.*

Under the current standards and recommended practices issued by ICAO, any medical assessment of the health status of aircraft maintenance engineers is not required at any stage, except for flight engineers (2012). Consequently, aviation safety authorities such as EASA place the subject in its very general and vague form, since they only require that the holder of a license or certificate issued under EASA Part-66 "must not exercise the privileges of that license or certificate while the person has a known medical deficiency, or increase of a known medical deficiency, that creates a risk of harm to that person or any other person." Additionally, in an effort to safeguard their legitimate interests, maintenance organizations usually require a basic medical assessment prior to recruitment, at their own level, which may include audiogram and vision tests.

Strategies to cope with stressors require the involvement and commitment of all stakeholders: the maintenance organizations, the engineers, the supervisors and the management. Nowadays, at the corporate level, the establishment of an effective *safety management system* (SMS) that is gradually reinforced via existing regulations constitutes the main underlying policy framework, which assists the management in monitoring safety in the workplace and dictates the need for proactive measures, in order to assess the likelihood of safety risks and their severity of impact. It is a pragmatic operational platform that takes into account the prevailing human factors and stressors by comprehensively incorporating various safety monitoring tools and on-the-job trainings, such as the already mentioned *fatigue risk management system,* the *aviation safety reporting system* and the *maintenance resource management training* that are further elaborated in Chapter 7.

On a personal level the successful management of work-related stressors can be achieved by training the engineers on the effects of stressors on their health and performance, as well as by their gaining confidence and familiarity with safety management systems. The ability to recognize the symptoms is also important so as to prevent the possibility of taking on high-risk tasks. Consequently, any relevant academic and practical training should begin at an early stage, within the training maintenance institutions and be repeated frequently in real working conditions. As a general mandatory rule, engineers should follow the manuals and checklists of the aircraft, thereby reducing the mental and physical demand, as well as the likelihood for memory lapses. Particular attention should also be paid when tasks are handed over to the next shift (Hobbs 2008), which implies that before a task is executed engineers must be confident that they are familiar with the procedures and that they have gathered all the necessary information and resources required. In addition, whenever they are forced to interrupt a task, in order to switch to a more urgent assignment, they should keep in mind that this change always involves risk, since it creates more workload and stress and can only be limited with task securement and recording on work cards.

Similarly, when supervisors allocate tasks they should consider the associated workload in relation to the physiology and mental condition of individuals, by looking for any visual symptoms of stress and fatigue. Furthermore, it is equally important that they allocate tasks according to their complexity and imposed time constraints, depending on the experience, knowledge and skills of engineers, whilst simultaneously monitoring their work by providing guidance and solutions.

### 4.6.1 INCIDENT/ACCIDENT EXAMPLE 9: THE CASE OF JET2.COM, JULY 23, 2011

During a landing roll at Paris Charles de Gaulle Airport in France, a B737–300 began to deviate slightly to the right of the runway centerline, after it suffered violent vibrations as soon as the wheels touched down, especially through the rudder pedals. The oscillations stopped when the aircraft decelerated at 75 kt, at a point when the aircraft had deviated about 10 m to the right. Then, the crew returned the aircraft onto the centerline, and came to a stop after they evacuated the runway and before being towed to the ramp. The aircraft suffered significant damage on the right main landing gear.

The investigation conducted by the French Office of Investigation and Analysis for the Safety of Civil Aviation (BEA 2012) revealed that the incident was the first flight after the aircraft had undergone maintenance the previous day that included the replacement of the landing gear. Furthermore, the investigation showed that the maintenance failed to reconnect the shimmy damper back to the hydraulic supply line, thus making it impossible for the actuator piston to absorb and counter the resulting vibrations during high rolling speeds.

At the time of the maintenance of the aircraft, the maintenance teams had to manage a second airplane that was in the hangar, while at the same time there was the need to repair a third aircraft that arrived with a problem on the fuselage. The prevailing conditions according to the report were the simultaneous needs to manage multiple tasks and the additional workload that derived from unplanned maintenance, as well as the excessive number of working hours of the supervisor, which exceeded the limited capacity of the maintenance teams and may have caused time pressure, high workload and fatigue. It is noteworthy that the task of reconnecting the shimmy damper to the hydraulic system was stamped on the work card although it was omitted, a fact which reveals the severity of the prevailing stressors on human cognitive processing by limiting attention and memory capabilities as well as decision-making.

### Lessons Learned #6
The stressors mentioned in this chapter are a real threat to the health of engineers and to aircraft safety, which is confirmed through empirical evidence in the workplace of maintenance, but also in the aftermath of incident reports and accident investigations. Time pressure, workload, shift work and the consequent symptoms of stress and fatigue can be managed through a systemic and comprehensive approach; however this requires education as well as training and commitment.

## REFERENCES

Adan, Ana, Simon N. Archer, Maria Paz Hidalgo, Lee Di Milia, Vincenzo Natale, and Christoph Randler. 2012. Circadian Typology: A Comprehensive Review. *Chronobiology International.* 29(9): 1153–1175.

Akerstedt, T. 1998. Shift Work and Disturbed Sleep/Wakefulness. *Sleep Medicine Reviews.* 2(2): 117–128.

BEA (Bureau d'Enquêtes et d'Analyses pour la Sécurité de l'Aviation Civile). *Vibrations, failure of the right main landing gear torsion link during landing roll,* released March 2012.

Bedell-Avers, Katrina E., William B. Johnson, Joy O. Banks, Darin Nei, and Elizabeth Hensley. 2011. *Fatigue Solutions for Maintenance: From Science to Workplace Reality.* Washington, D.C.: Federal Aviation.

Charlton, S. G. 2002. Measurement of Cognitive State in Test and Evaluation. In *Handbook of Human Factors Testing and Evaluation,* ed. S. G Charlton and T. G. O'Brien, 97–126. Mahwah, NJ: Lawrence Erlbaum.

Diamond, D. M., A. M. Campbell, C. R. Park, J. Halonen, and P. R. Zoladz. 2007. The Temporal Dynamics Model of Emotional Memory Processing: A Synthesis on the Neurobiological Basis of Stress-Induced Amnesia, Flashbulb and Traumatic Memories, and the Yerkes–Dodson Law. *Neural Plasticity.* 2007: 1–33.

Drake, Christopher L., and Kenneth P. Wright. 2011. Shift Work, Shift-Work Disorder, and Jet Lag. In *Principles and Practice of Sleep Medicine,* ed. M. H. Kryger, T. Roth, and W. C. Dement, 784–798. St. Louis, Missouri: Elsevier Saunders.

FAA (Federal Aviation Administration). 2012. *Advisory Circular 117–3.*

Francis Chaouloff, Sarah Dubreucq, Isabelle Matias, and Giovanni Marsicano. 2013. Physical Activity Feel-Good Effect. In *Routledge Handbook of Physical Activity and Mental Health,* ed. Panteleimon Ekkekakis, 71–87. Hoboken: Taylor and Francis.

Haidarimoghadam, R., R. Kazemi, M. Motamedzadeh, R. Golmohamadi, A. Soltanian, and M. R. Zoghipaydar. 2017. The Effects of Consecutive Night Shifts and Shift Length on Cognitive Performance and Sleepiness: A Field Study. *International Journal of Occupational Safety and Ergonomics: JOSE.* 23(2): 251–258.

Hebb, D. O. 1949. *The Organization of Behavior.* New York: Wiley & Sons.

Hobbs A. 2008. *An Overview of Human Factors in Aviation Maintenance.* ATSB Transport Safety Report. AR-2008–055. Australian Transport Safety Bureau.

Hobbs, A., Williamson, A., and Van Dongen, H. P. A. 2010. A Circadian Rhythm in Skill-Based Errors in Aviation Maintenance. *Chronobiology International.* 27: 1304–1316.

Hobbs, Alan, Katrina E. Bedell-Avers, and John J. Hiles. 2011. *Fatigue Risk Management in Aviation Maintenance: Current Best Practices and Potential Future Countermeasures.* Washington, D.C.: Federal Aviation Administration, Office of Aerospace Medicine.

Homan, W. J. 2002. Stress Coping Strategies for Commercial Flight Crewmembers. *Journal of Aviation Aerospace Education and Research.* 12: 15–28.

ICAO (International Civil Aviation Organization). 2012. *Manual of Civil Aviation Medicine.* Doc 8984.

Kobbeltvedt, Thérèse, Wibecke Brun, and Jon Christian Laberg. 2005. Cognitive Processes in Planning and Judgements under Sleep Deprivation and Time Pressure. *Organizational Behavior and Human Decision Processes.* 98(1): 1–14.

Leroy, Sophie. 2009. Why is it so Hard to Do My Work? The Challenge of Attention Residue when Switching between Work Tasks. *Organizational Behavior and Human Decision Processes.* 109(2): 168–181.

Lorist, Monicque M., Merel Klein, Sander Nieuwenhuis, Ritske De Jong, Gijsbertus Mulder, and Theo F. Meijman. 2000. Mental Fatigue and Task Control: Planning and Preparation. *Psychophysiology.* 37(5): 614–625.

Luckhaupt, Sara E., Martha A. Cohen, Jia Li, and Geoffrey M. Calvert. 2014. Prevalence of Obesity among U.S. Workers and Associations with Occupational Factors. *American Journal of Preventive Medicine.* 46(3): 237–248.

MacGregor, D. G. 1993. Time Pressure and Task Adaptation: Alternative Perspectives on Laboratory Studies. In *Time Pressure and Stress in Human Judgment and Decision Making*, ed. O. Svenson and A. J. Maule, 73–82. New York: Plenum Press.

Mark, G. M., and Andrew Paul Smith. 2008. Stress Models: A Review and Suggested New Direction. *Occupational Health Psychology, European Perspectives On Research, Education and Practice.* 3: 111–144.

Moore, Don A., and Elizabeth R. Tenney. 2012. Time Pressure, Performance, and Productivity. *Research on Managing Groups and Teams.* 15: 305–326.

Moray, N. P., and Nato Special Program Panel on Human Factors. 1977. *Mental Workload: Its Theory and Measurement.* Springer: reprinted 2013.

NTSB (National Transportation Safety Board). *Report No. AAR-03/02*, released August 5, 2003. Retrieved from Aviation Safety Network. Flight Safety Foundation.

Ordóñez, L. D., Lehman Benson, and Andrea Pittarello. 2015. Time-Pressure Perception and Decision Making. In *The Wiley-Blackwell Handbook of Judgment and Decision Making*, ed. Gideon Keren and George Wu, 519–542. Chichester: Wiley-Blackwell.

Phillips, R. O. 2014. *What is Fatigue and how does it Affect the Safety Performance of Human Transport Operators?* Institute of Transport Economics. Oslo: Norwegian Centre for Transport Research.

Puetz, T. W. 2007. Physical Activity and Feelings of Energy and Fatigue: Epidemiological Evidence. *Sports Medicine – Auckland.* 36(9): 767–780.

Saksvik, Ingvild B., Bjørn Bjorvatn, Hilde Hetland, Gro M. Sandal, and Ståle Pallesen. 2011. Individual Differences in Tolerance to Shift Work: A Systematic Review. *Sleep Medicine Reviews.* 15(4): 221–235.

Schreuder, E. J. A., and T. Mioch. 2011. The Effect of Time Pressure and Task Completion on the Occurrence of Cognitive Lockup. http://resolver.tudelft.nl/uuid: 1f17082a-b34b-494f-89b3-a5b967f77f68.

Smith, Andrew P., and Hugo N. Smith. 2017. Workload, Fatigue and Performance in the Rail Industry. In *H-WORKLOAD 2017*, ed Luca Longo and M. Chiara Leva, 251–263. Cham: Springer International.

Son, Joonwoo, Bruce Mehler, T. Lee, Y. Park, Joseph Coughlin, and Bryan Reimer. 2011. *Impact of Cognitive Workload on Physiological Arousal and Performance in Younger and Older Drivers*. Proceedings of the Sixth International Driving Symposium on Human Factors in Driver Assessment, Training and Vehicle Design. 87–94.

Suzuki, Takahiro, Terry L. von Thaden, and William D. Geibel. 2008. *Influence of Time Pressure on Aircraft Maintenance Errors*. Urbana, IL.: University of Illinois.

Van Dongen, H. P. 2006. Shift Work and Inter-Individual Differences in Sleep and Sleepiness. *Chronobiology International*. 23(6): 1139–1147.

Van Dongen, H. P., and Dinges, D. F. 2005. Sleep, Circadian Rhythms, and Psychomotor Vigilance. *Clinics in Sports Medicine*. 24: 237–249.

Vine, Samuel J., Liis Uiga, Aureliu Lavric, Lee J. Moore, Krasimira Tsaneva-Atanasova, and Mark R. Wilson. 2015. Individual Reactions to Stress Predict Performance during a Critical Aviation Incident. *Anxiety, Stress, and Coping*. 28(4): 467–477.

Wells, Alexander T. 1998. *Air Transportation: A Management Perspective*. Belmont, Calif: Wadsworth.

Wickens, Christopher D., Justin G. Hollands, Simon Banbury, and Raja Parasuraman. 2016. *Engineering Psychology and Human Performance*. London: Routledge.

Wright, K. P., J. T. Hull, and C. A. Czeisler. 2002. Relationship between Alertness, Performance, and Body Temperature in Humans. *American Journal of Physiology*. 283(6): R1370–1377.

Yerkes, Robert M., and John D. Dodson. 1908. The Relation of Strength of Stimulus to Rapidity of Habit Formation. *Journal of Comparative Neurology and Psychology*. 459–482.

# 5 Physical Environment, Tasks and Hazards in the Workplace

This chapter analyzes the particular features of the physical environment, as well as the prevailing technical conditions in the workplace of aircraft maintenance engineers. This analysis completes the mosaic of those factors that affect the psychology and physiology of the engineers and inevitably lead to precarious behaviors and errors, which affect their own health and safety and also jeopardize the safety of aircraft.

It is essential to include the nature of the tasks undertaken by the engineers as well as the associated hazards that exist in the workplace in this chapter, so as to achieve a higher level of understanding through their interaction. Finally, the main risk assessment methodologies for losing control of a hazard are also presented, in line with the predominant approaches to risk management in aviation, in order to provide the basic knowledge for the development of the necessary proactive strategies.

## 5.1 PHYSICAL ENVIRONMENT

The work environment can be thought of, simply, as the environment in which people work. As such, it is a very broad category that encompasses the physical setting (e.g. heat, equipment), characteristics of the job itself (e.g. workload, task complexity), broader organizational features (e.g. culture, history) and even aspects of the extra-organizational setting (e.g. local labor market conditions, industry sector, work-home relationships).

(Briner 2000)

The importance of a well-organized workplace is also noted by the World Health Organization which notes that "workplaces that promote mental health and support people with mental disorders are more likely to reduce absenteeism, increase productivity and benefit from associated economic gains" (2017). Human factors and ergonomics are vital elements of every working environment, while the organizational behavior significantly affects and is affected by the workplace (Brief and Weiss 2002). When assessing the ergonomic risk, employers must always consider both the physical and the psychological impact on their employers. It is a fact that

**FIGURE 5.1**    Indoor Aircraft Maintenance Facility. (Source: AdobeStock ©RenatSadykov – stock.adobe.com. Under Standard License AdobeStock_96513978.)

physical impact can be easily identified, while psychological impact is usually latent and is revealed only under specific circumstances (e.g. highly stressed periods, errors, etc.). Hazardous environments lead to unsafe/risky attitudes and then the error is inevitable.

Provisions should be in place in order to allow maintenance to be carried out in facilities where the working environment is ergonomically correct. Aircraft maintenance activities can be undertaken both indoors as in Figure 5.1, and outdoors as in Figure 5.2. In principle, indoor working environments should possess adequate levels of lighting and appropriate conditions allowing personnel to work in comfortable temperatures, whilst the facilities should be free of contamination and undue noise distraction. Nevertheless, maintenance activities may be undertaken not just outdoors, but in some cases in the most unprepared/disorganized locations and in ambiguous conditions. Therefore, it is important to acknowledge how an inappropriate working environment may impact the personal health or the safety of aircraft and to implement appropriate measures to mitigate these conditions.

Consequently, having realized the importance of the working environment, authorities have established a straightforward regulatory framework, based on the precondition that maintenance operators maintain appropriate facilities for their corresponding maintenance activities.

### 5.1.1 Noise and Fumes

The impact of sound on the health of engineers and on the safety of aircraft maintenance is discussed extensively in Chapter 2. As a rule, engineers must be well aware

**FIGURE 5.2** Outdoor Maintenance Workplace. (Source: AdobeStock. ©Druxa – stock. adobe.com. Under Standard License AdobeStock_195717606.)

that hearing loss starts in a very subtle manner. This is very clear if we consider a transient noise exposure, perhaps a static engine run-up that causes a ringing sensation in the ears. At first it clears quickly, but with repeated exposures the ringing sensation takes longer to clear, which implies that the first effect could easily be overlooked.

Workers in aircraft maintenance sectors, either in line or base maintenance, must wear ear protectors. Further to the adverse impact on personal health, investigations and continuous monitoring of safety management systems have shown that the higher the noise levels, the higher the probability for unsafe acts.

Airport ramps and aprons, as well as maintenance hangars and aircraft themselves, are hazardous in nature. Consequently, the health of individuals may be severely impacted in the short or long term, when employees are exposed to contaminated air due to fluids, chemical substances and aircraft exhaust. According to the European Agency for Safety and Health at Work there are several maintenance activities that involve exposure to hazardous substances. A great number of them can also be spotted in aviation maintenance activities as below:

- General cleaning activities
- Metal degreasing
- Painting
- Welding
- Maintenance activities that involve the use of lubricating products
- Dermal and inhalatory exposure to cooling fluids (alcohols, glycols), strong acids in batteries (sulphuric acid), asbestos in (old) brake linings or packings and fuel emissions.
- Coating activities
- Fumes due to the sweeping of hangar floors or the removal of cabin insulation.

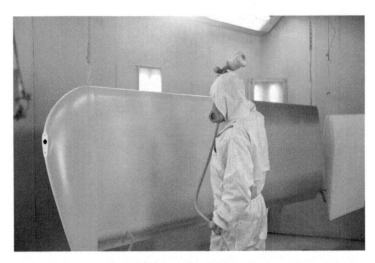

**FIGURE 5.3** Airplane Wing, Spray Painting, Worker in Coverall. (Source: AdobeStock. ©DenisProduction.com–stock.adobe.com. Under Standard License AdobeStock_183250704.)

When such activities take place, special precautions and preventative measures should be implemented (e.g Figure 5.3), such as hangar ventilation with fresh air at a rate of at least five to eight liters per second, per occupant. It is preferable for inlet air to also be filtered to remove particles. In cases where air conditioning is used, it should be regularly maintained in order to maintain purified and clean. On the other hand, when natural ventilation is used by opening windows it should not result in uncomfortable draughts.

### 5.1.2 ILLUMINATION

Having the limitations of human sight in mind that have been extensively discussed in Chapter 2, it is obvious that proper illumination in every aircraft activity is imperative, both for the quality of the work and the safety of the worker. Consequently, inadequate or damaged lighting should be replaced or maintained, otherwise it will constitute an unsafe latent condition and a hazard that should be identified, and risk assessed.

In outdoor work, natural light is the most suitable, since it incorporates the most optimal color rendition. Red-colored wire insulation, for example, will show up red in natural light, while if viewed under a mercury vapor lamp it will appear as orange or even black. However, if engineers need to work inside or after sunset, they have to use artificial lighting. Interestingly, the more efficient a light source is, the poorer it is at color rendition. Artificial lighting may be divided into the following basic categories:

- *Facility Lighting* that covers ordinary hangar and workshop fixed lighting systems. It should be evenly distributed without dazzling effects or any annoying glares.

- *Task Lighting* that is specialist lighting, fixed or adjustable or even mounted on the head of the individual and is used to provide better illumination for a particular task.

Intensity of light is measured in candela (SI system) or lux. As an example, for signs that identify the emergency exits a 50 lux light is required, as compared to the light of 5000 lux units that would be required for fine bench work. The light source should be positioned in such a manner that it does not shine into the eyes or produce glare from any reflective surfaces. The type and positioning of all lights and fittings is important, so as to not cause a hazard such as fire, radiation, electrical short circuits etc. Lights should not become obscured, and should be replaced, repaired and cleaned as necessary before the level of lighting becomes insufficient. For example, incandescent bulbs and fluorescent tubes result in good color rendition but have poor efficiency, while mercury vapor lamps result in poor color rendition but are more efficient.

### 5.1.3 CLIMATE AND TEMPERATURE

Achieving a balance in the climatic conditions of the workplace is one of the most difficult tasks to sustain proper working conditions, since climate and temperature conditions in particular, could affect employees in numerous ways. Maintaining temperature at appropriate levels is an important health factor in the workplace, as it can have a direct impact on productivity, while it may contribute to latent errors. Sufficient temperature control is required to keep workers comfortable and focused on the job. This not always an easy task in aviation maintenance environments, since the size of a hangar depends on the body of the aircraft hosted. For this purpose, it is often the case that the roof is constructed in such a manner so as to allow the complete tail level within the structure. As a result, the sizes of hangars vary. Normally, hangars that are being used for the maintenance of large commercial airplanes are bulky. Being such large spaces, hangars contain a large volume of air and are prone to high air change rates. Since the doors are often open, air infiltration can seriously disturb comfort levels. These adverse conditions are frequently reinforced by the fact that airports tend to be built in extremely exposed locations.

In warmer climates, air conditioning may have to be utilized, as well as a system to control relative humidity. In general, the higher the relative humidity is, the lower the temperature will be for a comfortable working environment, especially when the assigned tasks require more physical effort.

Furthermore, engineers often carry out maintenance outdoors where the climate conditions can become harsh. Apart from the visible and foreseeable risks that work may involve under extreme weather conditions, such as strong winds or snow/hail, it is important to understand that the combination of temperature and humidity also affects safety and health. Very few studies have been undertaken regarding thermal comfort in outdoor environments, compared to indoor environments, despite the fact that the importance of the former is increasingly recognized, with changing climate and increase of heat stress in cities (Honjo 2009). According to the same study, comfort zones may differ between outdoor and indoor areas. The repercussions may

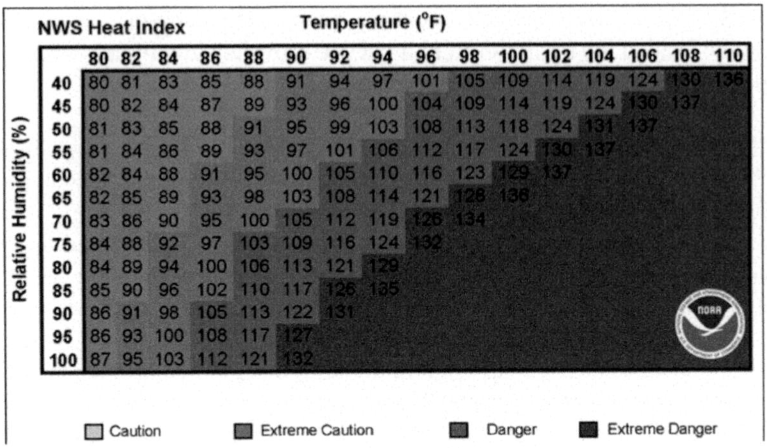

| NWS Heat Index | Temperature (°F) | | | | | | | | | | | | | | | |
|---|---|---|---|---|---|---|---|---|---|---|---|---|---|---|---|---|
| Relative Humidity (%) | 80 | 82 | 84 | 86 | 88 | 90 | 92 | 94 | 96 | 98 | 100 | 102 | 104 | 106 | 108 | 110 |
| 40 | 80 | 81 | 83 | 85 | 88 | 91 | 94 | 97 | 101 | 105 | 109 | 114 | 119 | 124 | 130 | 136 |
| 45 | 80 | 82 | 84 | 87 | 89 | 93 | 96 | 100 | 104 | 109 | 114 | 119 | 124 | 130 | 137 | |
| 50 | 81 | 83 | 85 | 88 | 91 | 95 | 99 | 103 | 108 | 113 | 118 | 124 | 131 | 137 | | |
| 55 | 81 | 84 | 86 | 89 | 93 | 97 | 101 | 106 | 112 | 117 | 124 | 130 | 137 | | | |
| 60 | 82 | 84 | 88 | 91 | 95 | 100 | 105 | 110 | 116 | 123 | 129 | 137 | | | | |
| 65 | 82 | 85 | 89 | 93 | 98 | 103 | 108 | 114 | 121 | 128 | 136 | | | | | |
| 70 | 83 | 86 | 90 | 95 | 100 | 105 | 112 | 119 | 126 | 134 | | | | | | |
| 75 | 84 | 88 | 92 | 97 | 103 | 109 | 116 | 124 | 132 | | | | | | | |
| 80 | 84 | 89 | 94 | 100 | 106 | 113 | 121 | 129 | | | | | | | | |
| 85 | 85 | 90 | 96 | 102 | 110 | 117 | 126 | 135 | | | | | | | | |
| 90 | 86 | 91 | 98 | 105 | 113 | 122 | 131 | | | | | | | | | |
| 95 | 86 | 93 | 100 | 108 | 117 | 127 | | | | | | | | | | |
| 100 | 87 | 95 | 103 | 112 | 121 | 132 | | | | | | | | | | |

☐ Caution    ■ Extreme Caution    ■ Danger    ■ Extreme Danger

**FIGURE 5.4** Likelihood of Heat Disorders with Prolonged Exposure or Strenuous Activity. (Source: US Government. In public domain.)

be more severe as concerns the psychological aspects when we interact in outdoor versus indoor environments. This is an important conclusion since line maintenance is performed outdoors under variable weather conditions. The United States Department of Labor refers to the heat index (Figure 5.4) as "the single value that derives from air temperature and relative humidity" and suggests that protecting measures should be taken by employers whenever the heat index exceeds permissible levels. The risk of heat-related illness becomes greater as the weather gets hotter and more humid.

### 5.1.4 MOTION AND VIBRATION

Although not common in aircraft maintenance, engineers should be aware of the danger of being exposed to specific frequency range vibrations, associated with the use of rotating or percussive tools. Mechanical vibrations at work can expose workers to hand-arm vibration (HAV) and or whole-body vibration (WBV).

Long-term regular exposure to HAV could result in potentially permanent and detrimental health effects, known as hand-arm vibrational syndrome, such as vibrational and carpal tunnel syndrome. On the other hand, regular long-term exposure to high levels of WBV is associated with back pain. The White Vibration Finger (VWF), also known as HAVS or Dead Finger Syndrome, is an industrial injury caused by the continuous use of handheld vibration that affects workers. It is a disorder that affects the blood vessels, nerves, muscles and joints of the hand, wrist and arm.

### 5.1.5 ERGONOMICS

Further to the above conditions, the working environment must be ergonomically correct, which means that workplaces should be designed or arranged in such a manner, so as to accommodate the people who use them. In general, the design of the working environment must contribute to the following goals:

- Fewer human errors
- Fewer health problems
- Fewer mental problems
- Fewer injuries or illness
- Increased productivity
- Higher quality.

In order to achieve the aforementioned goals, the parameters that should be considered when contemplating ergonomics intervention in the maintenance working environment, should be the following:

- Tools and technology
- Organization management
- Work processes
- Frequent tasks
- Workplace arrangement
- People.

### 5.1.5.1 Tools and Technology

The tools and technology used in any aviation maintenance environment should be ergonomically appropriate, in order to assist engineers in performing their tasks with safety and to prevent potential injuries or any damages to the aircraft. Platforms and any other supporting means should be appropriate and regularly maintained, in accordance with the manufacturers' instructions, thus enabling engineers to use them without fear of collapsing or damaging the aircraft. In addition, all serviceable tools and equipment should be in perfect working condition and calibrated. The toolboxes that the engineers carry should be ergonomically designed, in order to be as light as possible, while they should also allow engineers to easily identify any missing tools, in order to avoid using unsuitable tools that may harm aircraft systems.

Today's technologies allow engineers to collect real-time critical data concerning the operation of different aircraft systems and to implement prescriptive maintenance. In other words, not only can a fault be detected, but it can also be predicted to some extent in order to provide the optimal solution. In addition, through new technologies, engineers can contact the manufacturers directly to report problems and receive appropriate instructions, while they can improve the organization's spare parts supply chain management. It is clear, however, that the use of these technologies requires not only appropriate training, but also an assessment of the risk involved, as concerns the transition from old to new technologies.

### 5.1.5.2 Organization Management

The way that every organization is managed can be a predictor of its overall safety culture. Normally, the management should be able to protect workers and facilitate their work with all necessary means. However, whenever the feedback from the engineers requires additional means, it should be immediately examined and fulfilled. If the satisfaction of requests is not possible, then the reasons should be adequately explained and alternatives provided, otherwise they will become latent factors which will likely lead to bad norms.

As discussed in Chapters 3 and 4, the role of management and supervisors is crucial in aircraft maintenance, especially as concerns the composition of maintenance teams and identifying and eliminating the stressors created within them. Due care should be given to the newly hired personnel, who should receive all the necessary resources and guidance. Furthermore, the management should set authorized tasks to be accomplished within realistic timelines from an adequate workforce, in order to prevent time pressure and high workload that can lead to precarious shortcuts and errors.

### 5.1.5.3 Work Processes

Work processes refer to the actual work/task that maintenance workers should perform on the aircraft or parts of it. Despite the fact that aircraft manuals are very well-documented, the successful accomplishment of tasks within given time limits always depends on the experience and knowledge of the respective individual. It is imperative that all maintenance procedures are well-known and ordinarily applied by engineers. The engineers should perform all the tasks according to the maintenance manuals, avoiding any shortcuts, since the latter can easily lead to unsafe acts. Basic standard practices and processes should also apply, according to the relevant technical manuals, while the managers should confirm, systematically, that the maintenance personnel do not deviate from the proper processes. Depending on the task, there are work processes that vary in difficulty and complexity. It is also understandable that the accomplishment time of the tasks could differ, depending on the experience of the engineer, the conditions under which the task is taking place or any other factors that affect the performance of the personnel. However, it should also be clear that the work processes must be maintained at the same, high-quality level, regardless of the difficulty or complexity, or the particular conditions that the engineer is facing when performing the task.

### 5.1.5.4 Frequent Tasks

Frequent tasks are the ones that are frequently performed and are thus "very well-known" in a maintenance organization. According to empirical evidence, although the use of manuals is compulsory, many engineers do not follow them during the process but only consult them when they finish the work. This occurs since they take advantage of the cumulative experience acquired through the repetition of tasks and the consequent reduction of the required cognitive and physical effort. It is critical for work processes to be monitored by supervisors, who must look for signs of complacency and fatigue.

### 5.1.5.5 Workplace Arrangement

Ergonomic arrangement of the workplace is a key factor that generally increases productivity and the organization's competitiveness, whilst it also reduces costs.

Especially in the environment of aircraft maintenance, it mitigates the risks of injury and enhances safety culture. To this end, when considering workplace arrangement, human capabilities, limitations and needs must be kept in mind. Depending on the personnel manpower and the corresponding workstations, national regulations provide that workplaces must be equipped with the following facilities, the arrangement of which is up to the management:

- Changing into work/protective clothing
- Washing
- Rest rooms
- Eating
- Toilets (which should be kept clean and tidy).

### 5.1.5.6 People

People working in any position in a maintenance organization, spanning from the Accountable Manager to the WC cleaner, are the biggest assets of the organization. As previously underlined, particular attention must be paid to newcomers in order for them to quickly adapt to the working environment. Specifically, dedicated training should take place as soon as they join the organization, so as to inform them of the health and safety procedures and to allow them to become familiar with their new assignment. Concurrently, all the procedures should be written, whilst sufficient time should be allocated to workers to study and understand them.

Newcomers' feedback is always a good source for improvements, which could not have been identified beforehand by existing personnel, since long-term work in an organization often leads individuals to compromise even in inappropriate environments.

**Lessons Learned #7**

Consider the following seriously:

- In areas where conversation is difficult at 2 meters, there is a noise hazard and thus workers should always wear hearing protection such as ear plugs and ear muffs.
- Employees must wear hearing protection, when working in areas where the noise exceeds 85 dB, averaged over an eight-hour working period.
- Areas where the noise exceeds 90 dB, averaged over an-eight hour working period, will be designated as "ear protection zones." The ZONE should be marked by warning signs which comply with Industrial Regulations.
- All noise reduction and insulation arrangements such as screens, vibration insulation and reduction precautions, will be maintained in an efficient state of repair. The results of tests and maintenance shall be recorded by the manager responsible for the area.
- Personal breathing masks/hoods should be worn to ensure a clean air supply and to prevent damage to the lungs.

- Carbon monoxide is a poisonous gas that emanates from the incomplete combustion of fuel. It can be detected at varying levels in all smoke and fumes from burning carbonaceous substances. It is colorless, odorless and tasteless. Symptoms of poisoning from carbon monoxide include nausea, headaches and vomiting, and can lead to unconsciousness and death, depending on the concentration and time.
- For indoor work the temperature should be from 16°C to 21°C for non-physical work, but a minimum of 13°C is allowed for manual work.
- When using pneumatic tools, injuries can occur at frequencies between 5 and 2000 Hz, but the greatest risk for fingers is between 50 and 300 Hz. Preventative measures may include anti-vibration gloves and frequent breaks.

## 5.2 TASKS

Aviation maintenance entails different types of tasks that have to be successfully accomplished, in order to deliver an airworthy aircraft back to service. Human factors expertise is vital for the correct performance of these tasks. Since physical work, repetitive tasks, visual inspections and complex systems are prone to human error, special importance must be given to protect human health and aircraft safety. Thorough, accurate and feasible planning is therefore vital for the correct and timely execution of tasks, which should include the mustering of the necessary manpower, tools and spare parts and the related documentation before starting work.

### 5.2.1 PHYSICAL WORK

Aircraft maintenance incorporates a wide range of tasks that require increased physical endeavors and can, as discussed in the preceding chapters, lead to injuries and unsafe actions, either isolated or in combination with mental efforts and other stressors. Such tasks may include repair or lubrication of systems and parts, wheel replacements, servo actuator replacements, flying controls manual checks etc. These tasks require optimal physical condition, since beyond physical strength they are often performed on platforms, on the aircraft surfaces or even at extremely inaccessible locations (bays, tail booms, etc.).

Body posture (Figure 5.5) is also critical in order to avoid injuries when lifting heavy items. As a rule, the height of the vice or bench should be analogous to the *elbow height* of the person using the bench. For filing or machinery work, the workpiece should be 5 cm lower than elbow height, whilst during visual inspections the workpiece should be between elbow height and 25 cm above. The floor should be firm and level, but when standing for long periods, duckboards (wooden platforms) are advantageous. Unsuitable workbench height may result in musculoskeletal problems such as shoulders, neck or lower back injuries. It is necessary to adopt the most appropriate posture, especially when lifting an object with excessive weight. Where the removal or fitting of parts is not possible by a single engineer, then cooperation with other colleagues is imperative. Alternatively, the use of special lifting equipment must be sought, always in accordance with the technical instructions.

**FIGURE 5.5** Proper Lifting Technique. (Source: AdobeStock. © elenabsl com – stock. adobe.com. Under Standard License AdobeStock_ 144115184.)

Working in confined spaces for prolonged periods of time is also problematic. It is important for engineers to identify the point where their concentration and focus has been reduced (which can lead to unsafe acts), so as to decide on whether they should take a break or ask another engineer to continue the task.

Proper clothing and uniforms are also important factors that need to be considered before commencing work. Overalls should be a good fit and comfortable, especially when it is necessary to work in odd body positions. The amount and type of clothing depends on the temperature, the location (outdoors or indoors) and the nature of the task. Evidently, when more clothing is worn, the movements of the torso, arms and legs are restricted. Furthermore, boots or shoes should be non-slip, industrial heavy-duty, in order to avoid becoming damaged or penetrated by fuel or oil. Also, they should not generate any sparks, and avoid scratching aircraft structures or damaging the skin of the aircraft.

## 5.2.2 Repetitive Tasks

Tasks are classified as repetitive if the task duration is comparatively short and task frequency is high. For example, if an engineer is tasked to fit five rivets it could hardly be classified as a repetitive task, but if he/she had to put the same rivets in day after day for months on end, then that would constitute a repetitive task. Similarly, more burdensome tasks on aircraft maintenance can become repetitive. Maintenance planning documents are provided by aircraft manufacturers to describe the repetitive tasks that are required to maintain their aircraft. A study conducted in the manufacturing industries revealed that a total of 48.8 percent of the variance in human error can be explained by stress, repetition, fatigue and work environment (Yeow et al. 2014).

In general, repetitive tasks involve the use of motor memory, which implies that they can be undertaken with relative ease, and create complacency since they are done without putting in much "real thinking." Recalling the Yerkes–Dodson Curve of arousal, it is evident that task repetition may result in performance cutbacks caused by boredom due to monotony, which is detrimental to a worker's health, well-being and performance (Loukidou et al. 2009). As a result, supervisors should find ways to encounter boredom during work. To prevent boredom, psychologists suggest that short breaks and interchanging between tasks can be helpful. Thus, it is essential for supervisors to plan the allocation of tasks accordingly.

### 5.2.3 VISUAL INSPECTION

In aircraft maintenance, visual inspections comprise 80 percent of all inspection tasks, since they are quicker and more flexible, as compared to other inspection techniques. *Visual inspection* is described as "the process of using the unaided eye, alone or in conjunction with various aids, as the sensing mechanism from which judgments may be made about the condition of a unit to be inspected" (FAA 1997). It is perhaps best characterized as using the inspectors' senses, accompanied with only simple job aids such as magnifying loupes or mirrors (Drury and Watson 2002).

The physical environment as previously presented in this chapter plays a vital role for a successful visual inspection, since human capabilities should be adequate, in order to minimize hearing, vision and information processing limitations. Additionally, whenever engineers perform visual inspections as a part of a task, especially in areas that are not frequently visited, they must conduct a systematic visual search at a close distance to the observation area. This can be achieved by moving their eyes carefully in a set pattern, so that all parts are inspected and any abnormalities detected. Every panel removal provides a good opportunity to inspect the areas covered, especially for hidden problems like corrosion, or loose rivets or screws.

### 5.2.4 COMPLEX SYSTEMS

Complexity can be described as the system for which a single input has multiple outputs, thus making diagnostic and troubleshooting procedures very time consuming, mainly due to high levels of uncertainty for multiplex and integrated systems. The Federal Aviation Administration defines a complex airplane as an aircraft that has all of the following: a retractable landing gear, a controllable-pitch propeller and movable or adjustable flaps.

In principle, aircraft maintenance manuals are documented in such a manner so as to relax these complex systems and assign engineers many simple tasks to perform. It is critical when performing inspections or any other type of maintenance on an aircraft to follow the manuals and avoid performing any tasks without them, since as previously analyzed, complex tasks increase mental demand and impair information processing and *decision-making*. Having in mind that the false installation of components constitutes one of the biggest risks in aircraft safety, the associated tasks must be executed according to the documentation.

*Type rating* training is dedicated to one aircraft type; it aims to acquaint engineers as quickly as possible with the systems used, in order to mitigate complexity that often acts as a stressful factor, which multiplies as time pressure increases. Special attention should be given when incomplete tasks are handed over between shifts. To this end, the shifting teams should allocate enough time and resources to perform proper briefing in regards to the following:

- Confirmation that the job cards are signed and up-to-date
- Detailed verbal analysis of the progress of the job to date
- Indication of possible problems that may arise
- Report on the tools, workforce and spare parts availability
- Establishment of contact channels for liaison purposes.

## 5.3  HAZARDS IN THE WORKPLACE

Aviation is a sensitive and risk-prone industry, whereas the aircraft maintenance workplace incorporates numerous hazards that can harm personnel's health and impact aircraft safety. By definition, a hazard is any existing or potential condition that can impact the system and lead to: injury, illness or fatalities; damage to or loss of a system, equipment or property; or damage to the environment. A hazard becomes a threat whenever control over the hazard is lost. Consequently, hazard identification is a prerequisite to control safety risk. Thus, it entails the projected likelihood and severity of the consequence or outcome from an existing hazard.

### Lessons Learned #8

- Try to identify any hazards by walking around your workplace. Assess what could potentially harm employees' health or impact aircraft safety. Report your findings and request corrective measures. Monitor and provide feedback.
- Corrosion on wing attachments is a major hazard to flight safety because it can lead to a fatigue crack, resulting in loss of control.
- Naked electricity wires or an inappropriate refueling constitute hazards threatening both ground personnel and aircraft, because if left unattended, they can either cause electrocutions and cause serious injuries/death or develop into fires and damage the aircraft.

### 5.3.1  RECOGNIZING AND AVOIDING HAZARDS

A clear understanding of hazards and their related consequences is essential for the implementation of sound safety risk management (ICAO 2018). Hazard identification is the very first and fundamental stage of safety risk management processes, which must be implemented in every safety management system. According to ICAO, organizations can identify hazards by investigating the safety occurrences (reactive methodology) and by using the safety information that they collect from the various sources in their database (proactive methodology). The process should always

## TABLE 5.1
## Safety Risk Probability

| Likelihood | Meaning | Value |
|---|---|---|
| Frequent | Likely to occur many times (has occurred frequently) | 5 |
| Occasional | Likely to occur sometimes (has occurred infrequently) | 4 |
| Remote | Unlikely to occur, but possible (has occurred rarely) | 3 |
| Improbable | Very unlikely to occur (not known to have occurred) | 2 |
| Extremely improbable | Almost inconceivable that event will occur | 1 |

*Source:*  Adapted from ICAO SMM 2018. Under ICAO Standard Conditions for reproduction of ICAO materials.

*Note:*   This is an example only. The level of detail and complexity of tables and matrices should be adapted to the particular needs and complexities of each organization. It should also be noted that the organizations might include both qualitative and quantitative criteria.

## TABLE 5.2
## Example Safety Risk Severity Table

| Severity | Meaning | Value |
|---|---|---|
| Catastrophic | • Aircraft/equipment destroyed<br>• Multiple deaths | A |
| Hazardous | • A large reduction in safety margins, physical distress or a workload, such that operational personnel cannot be relied upon to perform their tasks accurately or completely<br>• Serious injury<br>• Major equipment damage | B |
| Major | • A significant reduction in safety margins, a reduction in the ability of operational personnel to cope with adverse operating conditions as a result of an increase in workload or as a result of conditions impairing their efficiency<br>• Serious incident<br>• Injury to persons | C |
| Minor | • Nuisance<br>• Operating limitations<br>• Use of emergency procedures<br>• Minor incident | D |
| Negligible | • Few consequences | E |

*Source:* Adapted from ICAO SMM 2018. Under ICAO Standard Conditions for reproduction of ICAO materials.

| Safety Risk | | Severity | | | | |
|---|---|---|---|---|---|---|
| Probability | | Catastrophic A | Hazardous B | Major C | Minor D | Negligible E |
| Frequent | 5 | 5A | 5B | 5C | 5D | 5E |
| Occasional | 4 | 4A | 4B | 4C | 4D | 4E |
| Remote | 3 | 3A | 3B | 3C | 3D | 3E |
| Improbable | 2 | 2A | 2B | 2C | 2D | 2E |
| Extremely improbable | 1 | 1A | 1B | 1C | 1D | 1E |

Note.— In determining the safety risk tolerability, the quality and reliability of the data used for the hazard identification and safety risk probability should be taken into consideration.

**FIGURE 5.6** Example Safety Risk Matrix. (Source: Adapted from ICAO SMM 2018. Under ICAO Standard Conditions for reproduction of ICAO materials.)

involve an analysis of the associated risk in terms of its likelihood and severity and result in an assessment of risk tolerability, which constitutes an indication of the need for further safety risk mitigation. Figure 5.6 presents an example of a Safety Risk Matrix, while Figure 5.7 an example of Safety Risk Tolerability.

### 5.3.1.1 The Bow-Tie Analysis
The Bow-Tie analysis developed in the offshore oil and gas sector in the 1990s could be used as a primary tool for the identification of hazards and the assessment of the associated risk. There are many different options to develop the analysis; however, as a rule, the diagram should resemble a bow-tie and include the point where control over a hazard is lost (the undesired state where it becomes a threat to the system) in the middle. It usually depicts the threats and the measures to control the hazard on the one side and the possible consequences of the hazard scenario on the other. The diagram could include the proactive (preventive) measures and the reactive

| Safety Risk Index Range | Safety Risk Description | Recommended Action |
|---|---|---|
| 5A, 5B, 5C, 4A, 4B, 3A | INTOLERABLE | Take immediate action to mitigate the risk or stop the activity. Perform priority safety risk mitigation to ensure additional or enhanced preventative controls are in place to bring down the safety risk index to tolerable. |
| 5D, 5E, 4C, 4D, 4E, 3B, 3C, 3D, 2A, 2B, 2C, 1A | TOLERABLE | Can be tolerated based on the safety risk mitigation. It may require management decision to accept the risk. |
| 3E, 2D, 2E, 1B, 1C, 1D, 1E | ACCEPTABLE | Acceptable as is. No further safety risk mitigation required. |

**FIGURE 5.7** Example of Safety Risk Tolerability. (Source: Adapted from ICAO SMM 2018. Under ICAO Standard Conditions for reproduction of ICAO materials.)

**TABLE 5.3**

**Common Hazards in Aircraft Maintenance Workplace and Safe Practices**

| Hazard | Potential Threat | Safe Practices |
|---|---|---|
| Electrical sources such (naked cables, faulty sockets) | Can cause electrocution resulting in serious or fatal injuries. Increase the risk of fire | All high-voltage sources are clearly identified and marked. Wear appropriate gloves |
| Hazardous and flammable materials (paint strippers, acid, fuel etc.) | Can create chronic respiratory problems and other serious illnesses. Flammable materials can cause fire and have catastrophic results | Always follow written instructions in regard to use and storage of hazardous and flammable materials. Use special breathing masks, protective clothing and safety glasses |
| Slippery surfaces | Can cause serious or fatal injuries and damage wing areas | Use protective clothing, safety helmets and anti-slip shoes |
| Working at heights (use of ladders, benches etc.) | Can cause serious or fatal injuries when falling. Increases stress and reduces situational awareness | Be tethered whenever required. Check suitability of surfaces on which you are standing |
| Confined space areas (fuel tanks, landing gears etc.) | Increase stress, reduce situational awareness, impair information processing and affect decision-making | Keep best possible communication with your team. Stop when needed by recording your work. Indicate where access is not possible. Make sure job is completed and do not assume anything |
| Inappropriate or poor lighting | Can cause poor situational awareness and result in serious or fatal injuries. Can affect visual inspections by impairing color rendition or depth vision | Use appropriate lighting for given task. Follow the checklist. |
| Environmental climate | High temperatures and high humidity levels or freezing and icing conditions affect arousal and impact performance by reducing comfort level | Use appropriate clothing. Reschedule tasks or take frequent breaks by recording work done |

| Vibrating tools | Repeated use can expose workers to hand-arm vibration (HAV) and or whole-body vibration (WBV) and potentially permanent health effects | Avoid prolonged use. Use supports wherever possible. Maintain proper body posture |
|---|---|---|
| Heavy machinery | If misused can cause musculoskeletal disorders or serious or fatal injuries. Failure to follow procedures can damage aircraft surfaces and systems | Use safety helmets. Keep correct body posture. Ask for help or use cranes |
| Noise | Can cause hearing damage or loss and also other psychological effects. Can also impose lack of communication, distraction, fatigue and loss of situational awareness | Use headsets or earplugs. Double check your work. When distracted go back at least three steps when commencing work |
| Vehicle movement (GPUs, forklifts etc.) | Breaching the prescribed vehicle movement lanes can result in serious or fatal injuries | When necessary wear reflective fluorescent jackets. Keep out of vehicle movement areas |
| Aircraft danger areas (engine inlets and outlets, propellers, sharp edges etc.) | Failure to keep out of aircraft danger areas can cause serious or fatal injuries | Know the aircraft danger zones. Do not enter unless clearance is given. Avoid wearing rings, necklaces etc., which can get caught on aircraft surfaces and cause injuries or amputations |
| Foreign object debris (FODs) | Screws, bolts etc. left off in an aircraft can affect proper functioning of engines, systems and control surfaces | Always measure your tools before and after work. Wear overalls, whose zippers must be closed |

*Source:* Created by Yiannakides and Sergiou 2019.

measures in case of loss of control. Depending on the nature and size of the system under consideration, the sources of hazards derive from the following (FAA 2017):

- Ambient environment (physical conditions, weather etc.)
- Equipment (hardware and software)
- External services (contract support, electric, telephone lines etc.)
- Human–machine interface
- Human operators
- Maintenance procedures
- Operating environment (airspace, air route design etc.)
- Operational procedures
- Organizational culture
- Organizational issues
- Policies/rules/regulations.

### 5.3.2 DEALING WITH EMERGENCIES

In such a complex and diverse environment workers must be aware of the actions immediately necessary in case of an emergency in the workplace. As a rule their reaction should be based on the following: *assess the situation, then secure the place and finally act.*

Maintenance organizations must have emergency response plans in place, which should include the actions undertaken by all personnel and the associated responsibilities of authorities in case of an emergency. Emergency planning will ensure that an orderly and efficient response to the situation is achieved, in a manner that will avoid further damage, injuries or loss of life. To this end, emergency planning should be regularly tested while all employees must be informed and trained accordingly.

### 5.3.3 COMMON HAZARDS AND SAFE PRACTICES

The process of addressing the hazards that may exist in the workplace requires at least the ability to recognize the hazard. Therefore, a non-exhaustive list of the most basic hazards (including some protective measures) is depicted in Table 5.3, which, if not addressed in time, pose a threat to the health and safety of workers as well as to the safety of aircraft. Note that the list does not constitute a checklist of safe practices, which should be based on National Authorities Regulations and safety standards.

## REFERENCES

Brief, Arthur, and Howard M. Weiss. 2002. Organizational Behavior: Affect in the Workplace. *Annual Review of Psychology.* 53: 279–307. 10.1146/annurev.psych.53.100901.135156.
Briner, R. B. 2000. Relationships between Work Environments, Psychological Environments and Psychological Well-Being. *Journal of Occupational Medicine*, 50(5): 299–303.
Drury, Colin G., and Jean Watson. 2002. *Good Practices in Visual Inspection.* Applied Ergonomics Group, FAA.

FAA (Federal Aviation Administration). 1997. *Visual Inspection for Aircraft*. Advisory Circular AC-43-204.

FAA (Federal Aviation Administration). 2017. *Safety Risk Management Policy*. Order 8040.4B.

Honjo, T. 2009. Thermal Comfort in Outdoor Environment. *Global Environmental Research*. 13: 43–48.

ICAO (International Civil Aviation Organization). 2018. *Safety Management Manual*. 4th ed. Doc. 9859. Canada.

Loukidou, Lia, John Loan-Clarke, and Kevin Daniels. 2009. Boredom in the Workplace: More than Monotonous Tasks. *International Journal of Management Reviews*. 11(4): 381–405.

WHO (World Health Organization). 2017. *Mental Health in the Workplace*. Information sheet.

Yeow, Jian Ai, Poh Kiat Ng, Khong Sin Tan, Tee Suan Chin, and Wei Yin Lim. 2014. Effects of Stress, Repetition, Fatigue and Work Environment on Human Error in Manufacturing Industries. *Journal of Applied Sciences*, 14(24): 3464–3471.

# 6 Communication

Research in the field of *interpersonal communication* is vast and particularly important, since it is widely accepted that employee performance is higher and safer in workplaces where there is good *communication*. Although the definitions assigned to *communication* vary, they converge to the view that interpersonal communication demands the existence of at least the sender and the receiver, who transmit information from the one to the other, in order to create common understanding (Keyton 2011). Psychologists suggest that *communication* can be verbal, thus expressed in words, and non-verbal, where it can be achieved through actions.

Early human factors programs showed that training in interpersonal communication skills could be improved when applied in human factors in aircraft maintenance (Patankar and Taylor 2007). Consequently, the understanding of communication properties and its implications for aircraft safety is critical, since lack of satisfactory communication skills can "decay intellectual and technical knowledge and may negatively affect the decision-making process in such a dynamic environment" (Dos Santos et al. 2014). Consequently it may result in misunderstandings, cost money, increase workload and, even worse, compromise safety (Civil Aviation Safety Authority Australia 2013).

Notwithstanding the fact that lack of communication is widely perceived as a predominant contributing factor to errors and that it occurs more frequently during the handover of ongoing tasks between shift teams or on the job between engineers, safety accidents/incident reports and studies provide evidence that lack of communication also appears between the various stages of the organization due to problematic communication flow (Virovac et al. 2017). In this case, the sender could be a unit or a department within the organization, such as the planning or the quality department/sector that disseminates written documents, which need to be adequately interpreted by the receivers, whose ability to decode the message depends on prior experiences and frames of reference (McShane and Von Glinow 2003).

As shown in Figure 6.1, the typical structure of aircraft maintenance organizations requires the establishment of appropriate communication channels that will facilitate flow of communication in the following possible directions:

- *Downward communication*: this top-down line starts from the manager and ends at the line engineer. It has to pass several managerial levels and requires each superior level to pass down the information with accuracy and precision. The passing downwards of some directive, communication or instruction implies temporary "storage" of that information in the mind or in the "in-tray." Careful consideration must be given on how this information is stored and displayed. Downward communications are usually "directives" to produce actions by subordinates, or policies and procedures that have to be followed.
- *Upward communication*: this is a bottom-up line from the workers to the managers. Through this line, engineers usually report results or genuine information, which may or may not need action. This upward flow of information should include the supervisors at each level, as soon as the information is available, otherwise the supervisors will develop feelings of loss of status due to a loss of control. An efficient communication system both up and down will create an atmosphere of cooperation and goodwill between employees and management.
- *Horizontal communication*: this is the flow of information exchanged between employees who are positioned on the same level in the hierarchy of the organization. It satisfies the need for coordinating and planning between the base maintenance and line maintenance departments or between the shift supervisors.

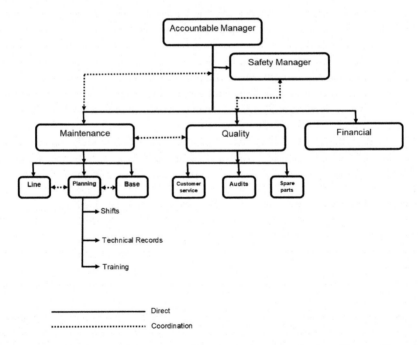

**FIGURE 6.1**  Typical Structure of Aircraft Maintenance Organization. (Source: Created by Yiannakides and Sergiou 2019.)

In aircraft maintenance, *communication* is achieved through written documents, symbols, diagrams, physical cues (Figure 6.2) and graphs, oral briefings and de-briefings, video teleconferences, telephone calls and e-mails. Communication can also be achieved by body expressions and signals that are analogous to the task, the working environment and the individualities of the people involved. ICAO 2003 suggests that *communication* can be *synchronous* or *asynchronous*. The first is of an informal nature and refers to verbal communication between engineers during briefings and task handling that is concluded with a minimal delay. On the other hand, *asynchronous communication* is formal, regulated and performed by work logging and recording; it includes job cards, reports, service bulletins and airworthiness directives, which are received or posted by engineers through electronic or physical means after a significant time lapse.

The usual shortfalls that impair effective *teamwork and communication* (Figure 6.3) are often related to unsent or incomplete messages, messages that are communicated through inappropriate channels and messages that are misinterpreted. Fortunately, *communication* skills can be improved with appropriate training and preparations including at least the following:

- Adequate planning of information and preparation to transmit the message through the appropriate communication modes and channels by taking into account possible obstacles
- Simplification of communication language to the level and knowledge of the receiver
- Acquisition of feedback in order to evaluate the effectiveness and efficiency of communication.

## 6.1 WITHIN AND BETWEEN TEAMS

Lack of or poor communication within or between teams can create "coordination problems that potentially render necessary safety procedures ineffective" (Suzuki et al. 2008). In the previous chapters, human body limitations, social psychology pressures and the physical environment impact have been demonstrated as the generating factors that can affect decision-making and communication processes. Stress and time pressure, for example, may negatively affect the sender's ability to prepare and transmit a concise message, while at the same time poor lighting conditions or elevated noise levels have the potential to reduce the comprehension of the receiver.

The different spoken languages among team members also impose a safety risk, due to the fact that maintenance engineers need to be capable to communicate and coordinate in the English language. This situation has been under the microscope of authorities and researchers for many years, since it affects their ability to understand and apply the aircraft manuals, which are documented in English. Researchers suggest that language barriers affect both synchronous and asynchronous communication, by imposing time pressure and difficulties in verbal communication and comprehension of written documents and safety placards (Drury and Ma 2003). Consequently, as a mitigating measure they recommend language training and the use of translation tools as appropriate solutions to alleviate the problem (Ma et al. 2010).

**FIGURE 6.2**   Physical Cue. (Source: AdobeStock. ©migrean – stock.adobe.com. Under Standard License AdobeStock_ 145321939.)

The successful transfer of information between the members of changing shifts, in the case of uncompleted tasks, depends on the turnover process and the communication skills of the individuals. To this end, it is critical that maintenance organizations establish and supervise the turnover procedures, which should include face-to-face briefings with mandatory work documentation signed by the engineers involved in the task work. The Federal Aviation Administration (FAA) puts it very clearly: "it is vital that work not be continued on a project without both oral and written communication between the technician who started the job and the

**FIGURE 6.3**   Effective Teamwork. (Source: AdobeStock. ©Gorodenkoff Productions OU. Under Standard License AdobeStock_ 218990793.)

technician continuing it" (US DOT 2018). The FAA also stresses the importance of team composition that is presented in Chapter 3 in this book. According to the recommendations of the FAA, effective teamwork should include the following ten characteristics: clear purpose, relaxed interaction, participation, listening, comfortableness with disagreement, openness, clear expectations, shared leadership and responsibility, good relations with others, team maintenance.

### 6.1.1 Incident/Accident Example 10: The Case of Continental Express, September 11, 1991

An Embraer 120 experienced loss of control due to the separation of the left horizontal stabilizer leading edge and crashed in Texas killing all people on board. According to the accident investigation report (NTSB 1992) the airplane had undergone a scheduled maintenance the night before the accident. In particular, among other tasks, two consecutive shifts were involved in the replacement of the two de-ice boots at the horizontal stabilizer. Investigators reported that lack of compliance by engineers and inspectors with the approved maintenance manual procedures led to the dispatch of a non-airworthy airplane, since 47 screw fasteners in the upper row of the left horizontal stabilizer leading edge were missing. The investigators determined that the reasons behind the errors were of a systemic nature and could not be attributed to a single failure or to an individual engineer. Consequently, they also concentrated on the issue of poor *communication* observed during the briefings, by incoming and outgoing supervisors and engineers of the two consecutive shifts, who were involved in the maintenance, and inadequate work logging and recording. The investigators statement included in the report is indicative of the prevailing circumstances:

> The Safety Board believes that the accident would most likely not have occurred if this supervisor had solicited a verbal shift turnover from the two mechanics he had assigned to remove the deice boots, had passed that information to the third shift supervisor, had completed the maintenance shift turnover form, and had ensured that the mechanics who had worked on the deice boots had filled out the M-602 work cards so that the third shift supervisor could have reviewed them.

## 6.2 WORK LOGGING AND RECORDING

European regulation clearly states that maintenance tasks shall be logged and recorded. Concurrently it entails that data concerning the ongoing airworthiness for every aircraft should always be available and maintained for a period of at least 36 months after the maintenance, while US regulation requires that records must be retained for at least two years. However, as previously explained in Chapter 3, it is the responsibility of aircraft engineers to log and record the basic details of the maintenance undertaken and to sign the certificate of release to service (CRS), which is the document permitting the aircraft to fly. Retaining data for a certain period of time is the legal responsibility of maintenance organizations, since their recovery in the event of accidents and incidents is obligatory. Moreover, recorded

data is requested by authorities during audits. This data includes the vital historical background of the completed work and therefore significant conclusions can be
drawn in terms of processes followed, the resources used, total labor hours etc.

In addition to being mandated by existing regulations, work documentation also
facilitates and safeguards that tasks are completed according to the procedures and
that nothing is omitted. Routine tasks and major component changes usually have
pre-printed work sheets, with the job laid down in stages for certification signatures.
Non-routine tasks, such as repair and rectification, rely on the engineer completing
the work sheets, as the job progresses in a logical and safe manner, with dupli-
cate inspections and tasks correctly called up as required by the assignment. In
addition, since technical work can only be based on approved publications that are
often updated or revised, great care should be taken to ensure that documentation
is complete and up-to-date. To this end, the aircraft's customer services department
must be contacted immediately in the event that errors are identified or suspected in
manuals, drawings or elsewhere. The recording work on the "Job Cards" and "Work
Sheets" shall be in accordance with the approved maintenance manuals, service
bulletins, airworthiness directives and notices.

### 6.2.1 INCIDENT/ACCIDENT EXAMPLE 11: THE CASE OF
### BRITISH MIDLANDS AIRWAYS, FEBRUARY 23, 1995

A Boeing 737–400 airplane suffered a loss of oil pressure on both engines after
taking off from East Midlands Airport in the UK. The aircraft diverted and landed
safely at Luton Airport. The investigation (AAIB 1996) revealed that the aircraft
had been subjected to borescope inspections on both engines during the preceding
night. Failure of maintenance to refit rotor drive covers resulted in the loss of almost
all the oil from both engines. The report identifies several human factors issues
that contributed to the incident, such as improper briefing between shift handover,
inadequate logging and recording, as well as poor judgment due to night shift working and interruptions. The Investigators' statement clearly underlines the prevailing
circumstances:

> Up to the time of the handover, the Line Engineer had intended to complete the tasks
> himself and he had not made a written statement or annotation on a work stage sheet
> to show where he had got to in the inspections.

However, he had to hand over the task to the night base maintenance controller, who
accepted the task on a verbal handover and was consequently not fully cognizant of
what had been completed and on the remaining task.

### 6.3 KEEPING UP TO DATE: DISSEMINATION
### OF INFORMATION

It is obligatory that aircraft engineers be kept up-to-date with the latest service bulletins, letters and amendments to manuals that are pertinent to their responsibilities
(e.g. Figure 6.4 and Figure 6.5). Apparently, accurate and timely *dissemination of*

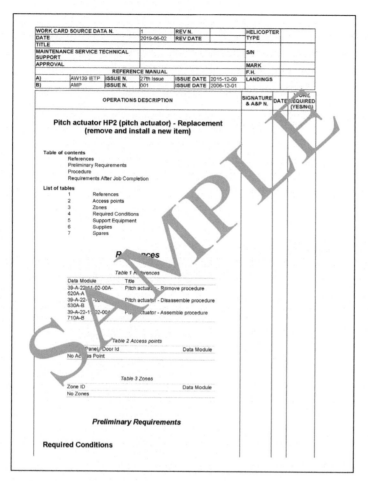

**FIGURE 6.4** Task Card Sample. (Source: Courtesy of Yiannakides and Sergiou's Archive.)

*communication* is critical to safety and implies the use of the appropriate channels. Consequently, changes in documentation should be disseminated as soon as possible and prior to the accomplishment of any relevant maintenance tasks, through safety notices on the company's notice boards and via the dissemination of copies of the amendments, so as to inform all parties involved in an efficient and timely manner. In order to ensure that the information has been received by the engineers, organizations usually require them to acknowledge receipt by signing. Concurrently, the European Regulation provides that "the person or organization maintaining an aircraft shall have access to and use only applicable current maintenance data in the performance of maintenance including modifications and repairs," whilst the United States Regulation stresses the need for updated documentation and assigns this responsibility to the quality control system.

It is also important to note that the continuous training of engineers on the latest developments and on the basic subjects covered within their initial training keeps them updated and enhances their competency. In order to reassure unhindered training, the European Regulation provides that "all staff involved in activities […] shall be appropriately qualified and have appropriate knowledge, experience, initial training and continuation training to perform their allocated tasks." Similarly, the US Federal Regulation requires that

> a certificated repair station must have and use an employee training program approved by the FAA that consists of initial and recurrent training […] the training program must ensure each employee assigned to perform maintenance, preventive maintenance, or alterations, and inspection functions is capable of performing the assigned task.

**FIGURE 6.5** EASA Airworthiness Directive Sample. (Source: EASA 2019. In public domain. Reproduced under EU Decision December 12, 2011 – Reuse of Commission Documents.)

**Lessons Learned #9**

Communication failure is an overarching human factor that leads to misunderstandings, erroneous decision-making and unsafe/hazardous acts. Based on the lessons learned, communication is a vital element of the fourth pillar of the safety management system "safety promotion," which aims at enhancing safety culture. This is achieved by highlighting the role of each engineer individually in safety and by optimizing team communication through continuous training on the inherent factors.

## REFERENCES

AAIB (Air Accidents Investigation Branch). UK Department of Transport. *Report No. AAR 3/96*, released June 26, 1996. Also in AAIB, *Report No. 3/1996*, published December 10, 2014.

Civil Aviation Safety Authority Australia. 2013. *Safety Behaviours: Human Factors for engineers*. Canberra.

Commission Regulation (EU) No. 1321/2014 of November 26, 2014 on the continuing airworthiness of aircraft and aeronautical products, parts and appliances, and on the approval of organizations and personnel involved in these tasks.

Dos Santos, Isabel Cristina, Ana Maria Vieira, and Paulo Renato de Morais. 2014. Poor Communication Skills Means High Risk for Aviation Safety. *Gestão & Regionalidade*. 30(88): 123–137.

Drury, Colin G., and Jiao Ma. 2003. Do Language Barriers Result in Aviation Maintenance Errors? *Proceedings of the Human Factors and Ergonomics Society Annual Meeting*. 47(1): 46–50.

ICAO (International Civil Aviation Organization). 2003. *Human Factors Guidelines for Aircraft Maintenance Manual*. Doc. 9824.

Keyton, Joann. 2011. *Communication and Organizational Culture: A Key to Understanding Work Experiences*. Los Angeles: SAGE.

Ma, Jiao, Colin Drury, and Clara Marin. 2010. Language Error in Aviation Maintenance: Quantifying the Issues and Interventions in Four World Regions. *The International Journal of Aviation Psychology*. 20(1): 25–47.

McShane, Steven Lattimore, and Mary Ann Young Von Glinow. 2003. *Organizational Behavior: Emerging Realities for the Workplace Revolution*. Boston: McGraw-Hill/Irwin.

NTSB (National Transportation Safety Board). *Report No. AAR-92/04*, released July 21, 1992. Retrieved from Aviation Safety Network. Flight Safety Foundation.

Patankar, Manoj S., and James C. Taylor. 2007. *Applied Human Factors in Aviation Maintenance*. Aldershot: Ashgate.

Suzuki, Takahiro, Terry L. von Thaden, and William D. Geibel. 2008. Coordination and Safety Behaviors in Commercial Aircraft Maintenance. *Proceedings of the Human Factors and Ergonomics Society Annual Meeting*. 52(1): 89–93.

US Department of Transportation. 2018. *Aviation Maintenance Technician Handbook – General*. FAA-H-8083-30A. Oklahoma.

US Federal Regulations CFR 14. Chapter I. Subchapter H. Part 145. Subpart E. §145.219.

Virovac, Darko, Anita Domitrović, and Ernest Bazijanac. 2017. The Influence of Human Factor in Aircraft Maintenance. *Promet – Traffic and Transportation*. 29(3): 257–266.

# 7 Human Error

Drawing on the conclusions of the preceding chapters, it is easy to conclude that *human error* emanates from latent behavior, due to individual or associated physiological, psychological, sociological and environmental factors, rather than an unforeseeable and accidental event. Obviously, not all errors belong to the same category, nor can their taxonomy be conducted with absolute certainty, based on measurable and unquestionable data.

ICAO defines error as "an action or inaction committed by an operational person that leads to deviations from organizational or the operational person's intentions or expectations" (2018). As will be underlined at a later stage, further to the above criterion, another condition that classifies a human error is the level of conscious control over the task and the pertaining behavior.

## 7.1 HUMAN FAILURE TYPES

Human error is the most frequent and common type of human failure that constitutes an undesired outcome, deviating from human intentions and expectations, whilst the alternative type of failure is classified as a *violation* (Rasmussen 1986; Reason 1990). *Errors* are unintentional actions and unplanned deviations that fail to generate the intended results, whereas violations are deliberate intentional transgressions that result in non-compliance with known rules, policies, procedures or acceptable norms.

*Errors* could be classified into three main types in accordance with the level of conscious control of the operator over the task: *slips*, *lapses* and *mistakes*. On the other hand, *violations* could be *routine*, *situational*, *optimizing* and *exceptional*.

In regards to the possibility to recover from a failure, researchers suggest that unintentional errors are more easy to identify and correct because the operator is usually more aware of the expected outcome of the task, while intentional mistakes are difficult to identify and recover because they tend to be resistant to disconfirming evidence (Embrey 2005).

### 7.1.1 BEHAVIOR OVER A TASK

The level of conscious control over a task is mainly determined by the behavior of the operators and their performance over a task, as well as from other indicators such as their training and experience and the nature of the task. Rasmussen suggests that human behavior over a task could be *skill-based*, *rule-based* or *knowledge-based* as follows (1986):

- *Skill-based behavior* during a task performance refers to the smooth execution of tasks without applying significant conscious control, based on the experience and skills of the operators, which have been acquired due to the repeated execution and practicing of the task on a routine basis.
- In a *rule-based behavior* the operator applies decision-making processing that has to be retrieved from memory, by applying optimal levels of conscious control. Rule-based behavior during a task performance refers to the execution of a task based on the memorization of rules and procedures that the operator has applied in previous occasions or has been taught to apply in a specific way.
- *Knowledge-based behavior* refers to complex or unfamiliar tasks undertaken by experienced operators that require significant levels of conscious control. Under these conditions the operator has no rules or routines available to handle the task and relies on principles and experience to accomplish the task. Knowledge-based errors concern decision-making and may reflect a lack of information on the task.

It is therefore obvious that the operator's initial purpose, and not the actual result of the error, is the fine line that distinguishes unintentional from intentional errors. However, aircraft accident investigations and legal processes have revealed that, sometimes, different types of errors coexist and unfold in a concurrent manner. Moreover, when a complex process is disrupted, individuals incorporate all three skill, rule and knowledge based performances, through continuous transitions in order to overcome the disturbance. Table 7.1 summarizes the various categories of human failures and associated types of errors and violations in an effort to make them more comprehensible.

## 7.2 ERROR MODELS AND THEORIES

Researchers have proposed numerous models and theories aiming to introduce a comprehensive approach, in order to predict or investigate unsafe acts in aviation, the most important of which are presented below.

### 7.2.1 SWISS-CHEESE MODEL

The "Swiss Cheese" model of accident causation (Figure 7.1) was originally proposed by James Reason and further developed by others (1990). This model enhanced safety thinking in aviation since it highlighted the significance of specific unsafe conditions that were not the most obvious during an investigation, usually due to the time and place of their occurrence.

The model is developed into a series of slices that replicates Swiss cheese. The slices represent the defenses against hazardous conditions within the various stages of the aviation system, from the design of the aircraft to its final preparation for flight. However, unsafe conditions, violations, errors or weaknesses inevitably breach the slices and create holes that can vary in size and are thus situated randomly in different positions. These holes are called *latent failures* and constitute the unsafe conditions that pre-exist in the aviation system, well before a damaging outcome is experienced. The actual effect of latent conditions may remain dormant for a long time and is usually not perceived as harmful, unless an *active failure* is experienced or a single error is generated by the personnel working in the front-line of aircraft operations such as engineers, pilots, air traffic controllers etc. The model suggests that whenever an active failure triggers a hazard that can impact the safety of the aircraft, this hazard has the dynamic to pass through the holes of the defenses of the system and momentarily align them, thus enabling the triggering of a chain effect that reinforces the hazard itself.

Moreover, latent conditions are generally created by engineers who have undertaken maintenance tasks at a past time or even possibly from previous workstations, whilst their current maintenance tasks may not be directly linked with the active failure or the specific time of the main event that led to the unsafe occurrence. Latent conditions may emerge from: a lack of safety culture; poor equipment or procedural design; conflicting organizational goals; defective organizational systems; or management decisions. Reason suggests that most accidents can be traced to one or more of four levels of failure:

- Organizational influences
- Unsafe supervision
- Preconditions for unsafe acts
- Unsafe acts.

The importance of the model lies in the fact that it addresses the latent and unsafe conditions within the causal sequence of events, thus enabling investigators and managers to consider them when reacting to accidents and when planning in a proactive manner, in order to mitigate their effect or if possible to prevent them from occurring in the first place.

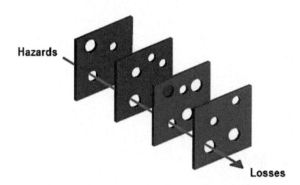

**FIGURE 7.1** Swiss Cheese Model. (Source: ICAO. Under Standard Conditions for Reproduction of ICAO Materials.)

**TABLE 7.1**

**Human Failures and Associated Types of Errors**

| Error Category | Unintentional | | | Intentional Non-Compliance | | | |
| --- | --- | --- | --- | --- | --- | --- | --- |
| | Error | | | Violation | | | |
| Type of Failure | Slip | Lapse | Mistake | Routine | Situational | Optimizing | Exceptional |
| Kind of unsafe act / Behavior over task | Skill-based associated with task execution | | Rule-based associated with task planning, decision-making and judgment / Knowledge-based associated with task planning, decision making and judgment | Rule-based and knowledge-based | | | |
| Associated level of conscious control | None or low | | Low to medium / Medium to high | Medium to high | High | | |
| Nature of tasks | Simple and routine task with known outcome / Visual inspections | | Simple and routine task with known outcome / Complex and unfamiliar task, unusual environment, unknown outcome | Routine teamwork tasks regardless of their complexity | Every task undertaken under specific conditions that include pressure, overload etc. imposed by organization or the environment | Every task undertaken for long time, under boredom or fatigue conditions | Every task undertaken under highly unusual conditions that make violations inevitable |

| | Attention diverted resulting in confusion, interference, reversal, misordering, mistiming | Memory shortage resulting in omissions, repetitions | Misapplication of good rule or application of bad rule | No availability of rules leads to incorrect diagnosis and application of inappropriate procedures | Following routine procedures outside the manuals. Perceiving existing rules as inadequate or time consuming | Undertaking risk without assessing outcome impact | Trying to impress others or to make the task more interesting | Taking unacceptable risk despite knowing the possible consequences |
|---|---|---|---|---|---|---|---|---|
| Immediate result | | | | | | | | |
| Probable contributing factors | • Complacency<br>• Stress<br>• Distraction<br>• Fatigue<br>• Lack of communication | | • Overload<br>• Time pressure<br>• Lack of knowledge<br>• Lack of awareness<br>• Peer pressure<br>• Stress<br>• Fatigue<br>• Lack of training<br>• Lack of resources | | | • Bad norms<br>• Peer pressure<br>• Time pressure<br>• Lack of assertiveness<br>• Fatigue<br>• Lack of teamwork | | |
| Mitigating measures | • Use check lists<br>• Avoid distractions and interruptions<br>• Cross-check your work<br>• Provide sufficient time<br>• Enhance situational awareness | | • Increase situational awareness<br>• Plan task<br>• Assess associate risk<br>• Provide/update procedures<br>• Improve competence | • Provide adequate training and resources<br>• Ensure proper supervision | | • Establish safety management system<br>• Promote safety culture<br>• Assess associated risk<br>• Do not take unacceptable risk<br>• Improve supervision<br>• Apply effective occurrences reporting system | | |

*Source:* Created by Yiannakides, Sergiou 2019.

### 7.2.2 SHELL MODEL

The "SHELL" model (Figure 7.2) analyzes the interaction of humans situated in the center of the system in relation to the various components of the system. It was first introduced by Elwyn Edwards in 1972 and further developed by Frank Hawkins in 1984 into a block structure (Hawkins and Orlady 1993).The name of the model originates from the initial letters of its components:

- *Software* (e.g. maintenance procedures, maintenance manuals, checklist layout, etc.)
- *Hardware* (e.g. tools, test equipment, the physical structure of aircraft, design of flight decks, positioning and operating sense of controls and instruments, etc.)
- *Environment* (e.g. physical environment such as conditions in the hangar, conditions on the line etc. and work environment such as work patterns, management structures, public perception of the industry, etc.)
- *Liveware* (i.e. the person or people, including maintenance engineers, supervisors, planners, managers, etc.).

The SHELL model is widely used in aviation since it takes into consideration the latent conditions that exist in the system, as well as the human factors capabilities and limitations when interacting with the system's components. According to the model, a mismatch between the Liveware and the other three components contributes to human error. Thus, these interactions must be assessed and considered in all sectors of the aviation system.

### 7.2.3 PEAR MODEL

The "PEAR" model (Figure 7.3) was proposed by William B. Johnson and Michael E. Maddox (2007). It is a memory aid model which has been applied for more than

**FIGURE 7.2**   SHELL Model. (Source: ICAO. Under Standard Conditions for Reproduction of ICAO Materials.)

| Physical Factors | Physiological Factors | Psychological Factors | Psychosocial Factors |
|---|---|---|---|
| • Physical size | • Nutritional factors | • Workload | • Interpersonal conflicts |
| • Gender | • Health | • Experience | |
| • Age | • Lifestyle | • Knowledge | • Personal loss |
| • Strength | • Fatigue | • Training | • Financial hardships |
| • Sensory limitations | • Chemical dependency | • Attitude | • Recent divorce |
| | | • Mental or emotional state | |

| Physical | Organizational |
|---|---|
| • Weather | • Personnel |
| • Location inside/outside | • Supervision |
| • Workspace | • Labor–management relations |
| • Shift | • Pressures |
| • Lighting | • Crew structure |
| • Sound level | • Size of company |
| • Safety | • Profitability |
| | • Morale |
| | • Corporate culture |

| | |
|---|---|
| • Steps to perform a task | • Knowledge requirements |
| • Sequence of activity | • Skill requirements |
| • Number of people involved | • Attitude requirements |
| • Communication requirements | • Certification requirements |
| • Information control requirements | • Inspection requirements |

| | |
|---|---|
| • Procedures/work cards | • Ground handling equipment |
| • Technical manuals | • Work stands and lifts |
| • Other people | • Fixtures |
| • Test equipment | • Materials |
| • Tools | • Task lighting |
| • Computers/software | • Training |
| • Paperwork/signoffs | • Quality systems |

**FIGURE 7.3** PEAR Model. (Source: Reproduced by permission of Dr. Bill Johnson (FAA).)

a decade and assists individuals to quickly recall the four most important human factors in aviation. The letter "P" stands for people work in maintenance; "E" represents the environment in which they work; the letter "A" recalls their actions; and finally "R" stands for the resources they use. This model is very useful and practical since it is straightforward, simply understood and easily memorized, despite the fact that it includes almost all the elements that may have an impact on safety.

## 7.3 ERRORS IN MAINTENANCE TASKS

Researchers, having exploited the available data from incident reporting systems, conclude that maintenance errors are strongly associated with the aircraft documentation, and in particular with the observed mismatch between the intention of the manual issuer and the consequent interpretation from the aircraft engineers (Chaparro et al. 2004) that results in procedural deviation (Lattanzio et al. 2008). The above factor reinforced by the various stressors that exist in aircraft maintenance, combined with the existent defective norms in organizations, facilitate the development of shortcuts in task accomplishment, which usually remain latent until they evolve into unsafe occurrences. Observations in workstations reveal that some engineers follow the documentation to the letter and step-by-step, once assigned a task, whilst others only consult the documentation when in doubt (Zafiharimalala et al. 2014).

Any maintenance task performed on an aircraft creates an opportunity for human error; however according to the CAA UK the most frequent errors in maintenance tasks are found as follows (2015):

- Incorrect installation of components
- Fitting of wrong parts
- Electrical wiring discrepancies to include crossing connections
- Forgotten tools and parts
- Inadequate lubrication
- Access panels, fairings or cowlings not secured
- Fuel or oil caps and fuel panels not secured
- Lock pins not removed.

### 7.3.1 INCIDENT/ACCIDENT EXAMPLE 12: THE CASE OF AIR CANADA, NOVEMBER 6, 2003

Following take-off from Vancouver on a scheduled flight to Calgary, an Airbus A330 was forced to return immediately to Vancouver, after the control tower informed the pilots that smoke was coming from one engine. After the pilots landed the airplane safely and none of the 92 passengers on board were injured, it was found that fuel was leaking from the engine, but there was no sign of fire.

The investigation conducted by the Transportation Safety Board of Canada (TSB 2004) revealed that the fuel leak was due to an error caused by maintenance technicians, in parallel with an error by the maintenance office duty board allocating tasks the previous day. In particular, during a routine service check, engineers observed a fuel leak from the engine involved in the accident. Consequently, maintenance ordered for a replacement of the air/oil heat exchanger by logging the task into the aircraft maintenance logbook. However, the maintenance office duty board mistakenly indicated on the task card that the task needed was the replacement of the fuel/oil heat exchanger. Although the maintenance team that undertook the task noted the discrepancies between the required and the allocated task, they replaced the fuel/oil heat exchanger without consulting the aircraft manual. This resulted in the omission of a retainer that is a crucial component securing the coupling, in order to restrict fuel leakage. In addition, in order to assure the quality of work, they performed an appropriate ground test for the replacement of the air/oil heat exchanger and not for the fuel/oil heat exchanger, thus minimizing any reasonable chances to observe the resulting fuel leak.

An in-depth analysis of this incident, in regards to the psychology of *error*, may indicate that that incident could be the result of the combination of a *slip* committed by the maintenance office duty board allocating tasks and a *mistake* from the maintenance team, in regards to the system installation and test. Particularly, the former could have committed a *slip* while displaying a skill-based behavior over a simple and routine task with a known outcome, over which it had no or very low levels of conscious control, probably due to distraction or complacency. Concurrently their attention was intermediately diverted, which resulted in confusion, reversal or misordering. On the other hand, the latter could have committed a *mistake* by having a rule-based behavior during a simple and routine task, with a known outcome that required decision-making and judgment, during which they had low to medium levels of conscious control. Despite the fact that the maintenance team did not perform the task according to the aircraft manual, probably due to the misapplication of a good rule or the application of a bad rule, *routine violation* may also have occurred, due to other human factors such as bad norms and time pressure.

*Again, it is neither the function nor the intention of the authors to assign fault or determine civil or criminal liabil ity as concerns any person involved in any incident cited in this book. Rather, the aim is to point out that the principles of human factors theory and the safety theoretical models apply to real events and consequently are crucial in preventing future accidents and incidents.*

### 7.3.2 INCIDENT/ACCIDENT EXAMPLE 13: THE CASE OF TUNINTER AIRLINES, AUGUST 6, 2005

An ATR-72 was forced to ditch into the Mediterranean Sea after both engines failed, due to fuel starvation, thus causing the death of 16 of the 39 people on board. The investigation conducted by the Italian National Agency for Flight Safety (ANSV 2006) revealed that the day before the accident the aircraft arrived at Tunis Airport with 790 kg of remaining fuel in the tanks to undergo maintenance. During the maintenance, engineers removed the fuel quantity indicator of the aircraft and installed a wrong indicator that was designed for the ATR-42 and not for the ATR-72. This error led to faulty higher fuel indications compared to the actual amount of fuel left in the fuel tanks. The installation error remained unnoticed by the maintenance, while the crew failed to detect the fuel exhaustion in regards to normal fuel consumption in the following two flights, including the last leg from Bari to Djerba during which the aircraft ditched.

ANSV presented an analysis of the various factors that contributed to the event by applying Reason's model. According to the analysis, the active failures that triggered the accident were committed both by ground mechanics/technicians and the crew. On the other hand, the latent failures were associated with the inadequate spares configuration management and deficient procedures for aircraft maintenance and technical management, which caused confusion and lack of clear roles and responsibilities of engineers. It should also be mentioned that, according to the investigation, the maintenance personnel had not attended any training in human factors prior to the accident.

## 7.4 IMPLICATIONS OF ERRORS

Errors in aircraft maintenance may lead to physical or mental harm and also jeopardize aircraft safety. In the framework of aircraft maintenance, errors typically emerge and are identified as follows:

- Immediately identified and rectified before the aircraft is authorized for its next flight
- Captured after the aircraft was authorized to fly and before take-off

- Not detected at all and manifested during flight
- Captured at an unknown time accidentally during a scheduled maintenance or inspection of the aircraft.

Consequently, the implications may be complex in nature and depend on the time
and phase of flight within which they are captured and on the manner in which
they develop. Even though they are not communicable to passengers, maintenance errors affect the public's trust as concerns the particular airline or their
confidence in air transportation in general, since they lead to delays or cancella-
tions of scheduled flights. In the worst case scenario, they cause injuries, fatalities and aircraft loss, or in many cases lead to the bankruptcy of maintenance
organizations and airlines.

Negative outcomes include heightened workload and occupational stress
since the time required correcting the error further increases time pressure.
Also, errors disrupt the consistency of maintenance teams and affect the individual's competency. In addition, errors question the organization's legitimacy
and impact stakeholders' support. Consequently, the implications of errors significantly affect not only the economic field, which in most cases can be predicted to a certain extent and addressed, but they also extend to workers' and
organizations' psychology.

In particular, as regards the impact of errors on cost burden, there are two types
of costs as follows:

- *Direct costs* that are explicitly visible, easy to determine and have the
  potential to be reduced by insurance coverage. They usually include com-
  pensations of aircraft major systems and components and sometimes the
  replacement of the aircraft itself. In fatal or serious accidents they can also
  include compensation for the loss of lives, injuries and post-accident medical expenses.
- *Indirect costs* that are partially covered or not covered at all by insurance
  companies. A part of these costs can also be hidden and amount to as
  much as four to six times greater than direct costs. Airlines have reported
  that more than 95 percent of such costs claims are below deductible and
  therefore uninsured. Indirect costs can derive from the main consequences
  below (Snyder and Ullrich 2016):
  - Disruption of business continuity due to shortfalls in equipment and
    human resources, which lead to flight delays or cancellations as well
    as to alterations of flight schedules
  - Impact on worker's productivity
  - Impact on business reputation and legitimacy
  - Legal actions, fines and citations taken against the company
  - Insurance deductibles
  - Accident investigation costs
  - Corrective measures imposed by authorities.

## 7.5 AVOIDING AND MANAGING ERRORS

Error management and avoidance must remain a top safety priority for the aviation community. ICAO suggests that organizations must put safety strategies in place, in order to control and eliminate errors in the workplace as follows:

- *Reduction strategies* in order to reduce or eliminate the factors contributing to the error. An example could be a strategy to reduce environmental distractions.
- *Capturing strategies* that intend to capture error since they assume the error will occur. Capturing strategies utilize checklists and other procedural interventions rather than directly eliminating the error.
- *Tolerance strategies* in an effort to increase the tolerance/resilience of the system and decrease the consequences since they assume that the error will be made. An example is the incorporation of redundant systems or multiple inspection processes.

Concurrently, the systematic training of employees and the establishment of an effective safety occurrences reporting system should be in the core of the existing and future strategies. Furthermore, nowadays, maintenance organizations have additional tools at their disposal that are broadly applied in aviation safety and can be utilized to achieve better safety performance standards: the *maintenance resource management* (MRM), the *safety management system* (SMS) and the Dirty Dozen.

### 7.5.1 AVIATION SAFETY REPORTING SYSTEM

Taking into consideration that accidents are preceded by safety-related incidents and deficiencies, which reveal the existence of safety hazards and unsafe behaviors, the application of effective proactive strategies requires organizations to establish occurrence reporting systems. To this end, existing global Regulations (Regulation (EU) No. 376/2014; Commission Implementing Regulation (EU) 2015/1018 of June 29, 2015; Designation of ASAP Information 2003) oblige maintenance organizations to establish mandatory and voluntary reporting systems, through which employees, and especially front-line personnel, have the ability to report occurrences that pose a significant risk to aviation safety. The list includes any occurrences related to technical and working conditions, as well as the maintenance and repair of the aircraft.

Reporting systems as EASA's in Figure 7.4, and NASA's in Figure 7.5 enable maintenance organizations, authorities and safety agencies to report, collect, store, protect, exchange, disseminate and analyze the incoming data and to build and monitor their safety strategies, by adopting immediate proactive and reactive measures. Since the successful application of the reporting systems depends on the trust and confidence shown by engineers in these tools, the applicable Regulations assist in promoting a *Just Culture* within maintenance organizations, which means that engineers are free to report and will not receive punishment. However, it must be clearly understood by all involved in aircraft maintenance that *Just Culture* does not

## AVIATION SAFETY REPORTING

**When and where**

| | | | |
|---|---|---|---|
| UTC Date - Time | - | Location of occurrence | |
| Local Date - Time | - | World region | |
| *(YYYY/MM/DD)* | *(HH:MM)* | State/area | |

**What**

Headline

Narrative language

Narrative

**Severity**

| | |
|---|---|
| Injury level | Highest damage |
| | *(to aircraft)* |

**Aircraft information**

| | |
|---|---|
| State of registry | Aircraft category |
| Aircraft registration | Manufacturer |
| Serial number | Model |
| Year built | *Other (specify)* |

**Flight details**

| | *Country* | *ICAO code* | *Other (specify)* |
|---|---|---|---|
| Last departure point | | | |
| Planned destination | | | |
| Operation type | | | |
| Flight phase | | Occ. on ground | |

**Operational information**

| | | |
|---|---|---|
| Weather relevant | Airspace class | Current flight rules |
| Weather conditions | | |

**Form: General Aviation on Personal Behalf**

Version 2.5 - August 2018 - AE 4.1.0.3                    1

Use Adobe Reader DC © to compile this form

Submit this form on www.aviationreporting.eu

**FIGURE 7.4** EASA Aviation Safety Reporting Online Platform. (Source: www.easa. europa.eu. In Public Domain. Reproduced under EU Decision December 12, 2011 – Reuse of Commission Documents.)

provide immunity for errors or unsafe behaviors, due to *violations,* the legal interpretation of which always remains the responsibility of the national judiciary (van Dam and Licu 2012).

## 7.5.2 MAINTENANCE RESOURCE MANAGEMENT (MRM)

The *Maintenance Resource Management* (MRM) was introduced by the aviation industry in the 1990s based on the *Crew Resource Management* training that commenced earlier. Following many developments, nowadays MRM training puts greater emphasis on teamwork and the pertaining concerns. Consequently, it

**D**

DO NOT REPORT AIRCRAFT ACCIDENTS AND CRIMINAL ACTIVITIES ON THIS FORM.
ACCIDENTS AND CRIMINAL ACTIVITIES ARE NOT INCLUDED IN THE ASRS PROGRAM AND SHOULD NOT BE SUBMITTED TO NASA.
ALL IDENTITIES CONTAINED IN THIS REPORT WILL BE REMOVED TO ASSURE COMPLETE REPORTER ANONYMITY.

(SPACE BELOW RESERVED FOR ASRS DATE/TIME STAMP)

**IDENTIFICATION STRIP:** Please fill in all blanks to ensure return of strip.
NO RECORD WILL BE KEPT OF YOUR IDENTITY. This section will be returned to you.

**TELEPHONE NUMBERS** where we may reach you for further
details of this occurrence:

**HOME**  Area _____ No. _____  Hours _____
**WORK**  Area _____ No. _____  Hours _____

**TYPE OF EVENT/SITUATION**
_____
_____

NAME _____

ADDRESS/PO BOX _____

_____
**DATE OF OCCURRENCE** _____
(MM/DD/YYYY)
CITY _____ STATE _____ ZIP _____  **LOCAL TIME (24 hr. clock)** _____
(HH:MM)

PLEASE FILL IN APPROPRIATE SPACES AND CHECK ALL ITEMS WHICH APPLY TO THIS EVENT OR SITUATION.

**EXPERIENCE**

| | |
|---|---|
| Describe your qualifications | ☐A  ☐P  ☐NDT  ☐repairman  ☐inspection authority  ☐avionics  ☐other _____ |
| What is your technician/maintenance experience in years? | lead technician _____   technician _____   repairman _____   avionics _____<br>inspector _____   other _____ |

**FACTORS**

| | |
|---|---|
| Location | _____ |
| Was training a factor? | ○ Yes   ○ No   Reset   ☐ I was instructing   ☐ I was receiving training |
| What other factors may have contributed? | ☐ lighting   ☐ work cards   ☐ briefing<br>☐ weather   ☐ manuals   ☐ other_____ |
| Check items which were involved in the event | inspection ○Yes ○No   installation ○Yes ○No<br>testing ○Yes ○No   scheduled maintenance ○Yes ○No   Reset<br>repair ○Yes ○No   MEL ○Yes ○No<br>logbook entry ○Yes ○No   *other _____<br>fault isolation ○Yes ○No   (*Describe in the Describe Event/Situation sector) |

Component/System/Sub-system involved: _____

| | | |
|---|---|---|
| Was maintenance deferred? ○Yes ○No<br>Reset | When was problem detected? | ☐ routine inspection   ☐ while aircraft was in<br>☐ in-flight        service at gate<br>☐ taxi   ☐ pre-flight<br>☐ other_____ |

**CONSEQUENCES/OUTCOME**

| | | | |
|---|---|---|---|
| ☐ flight delay<br>☐ flight cancellation | ☐ gate return<br>☐ air turn back | ☐ improper service<br>☐ rework | ☐ in-flight shut down<br>☐ aircraft/engine damage<br>☐ other_____ |

| **AIRCRAFT/AIRWORTHINESS STATUS** | **MISSION** | **REPORTER ORGANIZATION**<br>(Check all that apply) |
|---|---|---|
| ☐ aircraft released for service<br>☐ aircraft records completed<br>☐ aircraft required documents aboard<br>☐ not released for service<br>☐ unknown | ☐ passenger<br>☐ personal<br>☐ cargo/freight<br>☐ training<br>☐ ferry<br>☐ other_____ | ☐ air carrier     ☐ FBO<br>☐ air taxi      ☐ government<br>☐ contracted service  ☐ military<br>☐ corporate     ☐ personal<br>☐ fractional     ☐ other_____ |

**TYPE OF AIRCRAFT (MAKE/MODEL) AND ENGINE TYPE**

| | | | |
|---|---|---|---|
| type of aircraft _____ | series _____ | ATA Code _____ | Ê |
| aircraft zone _____ | engine model _____ | other _____ | Ê |

NASA ARC 277D (May 2009)   **MAINTENANCE FORM**   Page 1 of 3

**FIGURE 7.5**  National Aeronautics and Space Administration (NASA) ASRS Maintenance Report Form. (Source: Designation of Aviation Safety Action Program (ASAP) Information as Protected from Public Disclosure under 14 CFR Part 193. 2003. FAA-USA. *Aviation Safety Reporting System*. ASRS. National Aeronautics and Space Administration (NASA). Reproduced under NASA Media Usage Guidelines.)

enhances *safety culture* by providing engineers with the knowledge of how to assess safety attitudes and working methodologies and improve team communication and decision-making. It also incorporates the basic principles of human factors in error theory, by contributing to the development of *stress* management and *fatigue* strategies in particular.

### 7.5.3 Safety Management System

Nowadays, top-level management is developing safety models in order to comply with regulations and ensure safety in operations. The dominant model is the safety management system (SMS), which is a top-down approach and a decision-making tool, utilized and applied by all stakeholders including managers, supervisors and workers. SMS enables managers to assess the risks, predict probable safety occurrences and prioritize them, in terms of severity and probability. It is a tool that drives the decision-making process and dictates the proactive remedial actions and immediate reactive measures that should be implemented, in order to monitor safety risks and unsafe occurrences.

SMS transforms the organization's safety program into a functioning and viable safety management system, only if its four components work in harmony, which according to ICAO are safety policy, safety risk management, safety assurance and safety promotion (2013). A well-structured SMS provides organizations with the following advantages (Snyder and Ullrich 2016):

- A structured means of safety risk management decision-making
- A means of demonstrating safety management capability before system failures occur
- Increased confidence in risk controls through structured safety assurance processes
- An effective interface for knowledge sharing between regulators and operators
- A safety promotion framework to support a *safety culture*.

### 7.5.4 Dirty Dozen

In the aftermath of a series of maintenance errors in the late 1980s, Transport Canada with the support of the aviation industry identified the 12 most frequent human factors issues that affect engineers' safe performance and result in errors (see also Table 7.1). These factors are widely known as the *Dirty Dozen* and have become the cornerstone in human factors training worldwide, since they are easily comprehended and memorized, whilst they eventually assist aircraft engineers in enhancing awareness over potential error traps (Dupont 1997; Canada Transport Canada, UK Civil Aviation Authority and US FAA 1998). Although no factor is considered to be more important than others, often both professional engineers and students attach a particular weight to each of them. Nevertheless, Dirty Dozen is still a very useful tool to reduce errors and consequent accidents in the hangar and workshop maintenance (Samad et al. 2018).

- Lack of communication
- Complacency
- Lack of knowledge
- Distraction
- Lack of teamwork
- Fatigue
- Lack of resources
- Pressure
- Lack of assertiveness
- Stress
- Lack of awareness
- Norms.

### 7.5.5 European Strategy for Human Factors in Aviation

The European Union (EASA 2012) inaugurated a strategy for human factors in aviation with the aim to integrate human factors in the day-to-day operations of aviation organizations and thus improve safety (Figure 7.6) and effectiveness. Concurrently it serves as the framework for the European plan for aviation safety (EASA 2017) that considers human factors training as a strategic priority.

**FIGURE 7.6** Make Flying Safer. (Source: European Commission, Mobility and Transport. Under EU Decision December 12, 2011 – Reuse of Commission Documents.)

The strategy sets the foundations for training all employees in the critical aviation sector, as concerns the impact of human factors on aviation safety and thus requests the standardization of oversight standards and methodologies in error management. Finally, it promotes safety occurrences data reporting, collection and analysis by encouraging research and industry feedback.

### Lessons Learned #10

Human error in aircraft maintenance emanates from latent behavior, due to individual or associated physiological, psychological, sociological and environmental factors. It is not an unforeseeable and accidental event, but it can be predicted and mitigated through education and training.

## REFERENCES

ANSV (Agenzia Nazionale per la Sicurezza del Volo), Italy. *Final Report*, released 2006 (English version 2008).

CAA (Civil Aviation Authority) UK. 2015. *Aircraft Maintenance Incident Analysis*. CAP 1367.

Canada Transport Canada, UK Civil Aviation Authority and US FAA. 1998. *Symposium on Human Factors in Aviation Maintenance. The 12th Symposium on Human Factors in Aviation Maintenance*.

Chaparro, Alex, Bonnie Rogers, Chris Hambllin, and Barbara Chaparro. 2004. *Final Report: A Comparison of Three Evaluative Techniques for Validating Maintenance Documentation*. Embry-Riddle Aeronautical University.

Commission Implementing Regulation (EU) 2015/1018 of June 29, 2015, laying down a list classifying occurrences in civil aviation to be mandatorily reported, according to Regulation (EU) No. 376/2014 of the European Parliament and of the Council.

Designation of Aviation Safety Action Program (ASAP) Information as Protected from Public Disclosure under 14 CFR Part 193. 2003. FAA-USA. *Aviation Safety Reporting System*. ASRS. National Aeronautics and Space Administration (NASA).

Dupont, G. 1997. The Dirty Dozen Errors in Aviation Maintenance. In *Meeting Proceedings of Eleventh Federal Aviation Administration Meeting on Human Factors Issues in Aircraft Maintenance and Inspection: Human Error in Aviation Maintenance*, 45–49. Washington, D.C.: Federal Aviation Administration/Office of Aviation Medicine.

EASA (European Aviation Safety Agency). 2012. *European Strategy for Human Factors in Aviation*.

EASA (European Aviation Safety Agency). 2017. *European Plan for Aviation Safety 2018–2022*.

Embrey, David. 2005. *Understanding Human Behaviour and Error*. Dalton, UK: Human Reliability Associates.

Hawkins, Frank H., and Harry W. Orlady. 1993. *Human Factors in Flight*. Aldershot, UK: Ashgate.

ICAO (International Civil Aviation Organization). 2013. *Safety Management – Annex 19*. Montreal, Canada.

ICAO (International Civil Aviation Organization). 2018. *Safety Management Manual*. 4th ed. Doc. 9859. Canada.

Johnson, W. B., and M. E. Maddox. 2007. A PEAR Shaped Model for Better Human Factors. *The Journal for Civil Aviation Training*. 2: 20–21.

Lattanzio, Diane, Kirsten Patankar, and Barbara G. Kanki. 2008. Procedural Error in Maintenance: A Review of Research and Methods. *The International Journal of Aviation Psychology*. 18(1): 17–29.

Rasmussen, Jens. 1986. *Information Processing and Human-Machine Interaction: An Approach to Cognitive Engineering.* New York: North-Holland.

Reason, James. 1990. *Human Error.* New York: Cambridge University Press.

Regulation (EU) No. 376/2014 of the European Parliament and of the Council of April 3, 2014, on the reporting, analysis and follow-up of occurrences in civil aviation.

Samad, A. G. A., M. K. Johari, and S. Omar. 2018. Preventing Human Error at an Approved Training Organization Using Dirty Dozen. *International Journal of Engineering and Technology* (UAE). 7(4): 71–73.

Snyder, Paul R., and Gary M. Ullrich. 2016. *Practical Safety Management Systems: A Practical Guide to Transform Your Safety Program into a Functioning Safety Management System.* Newcastle: Washington Aviation Supplies and Academics.

Transportation Safety Board of Canada. *Report Number A03P0332,* released September 7, 2004.

van Dam, Roderick, and Tony Licu. 2012. *The Essence of Just Culture.* Eurocontrol.

Zafiharimalala, Herimanana, David Robin, and André Tricot. 2014. Why Aircraft Maintenance Technicians Sometimes do Not Use Their Maintenance Documents: Towards a New Qualitative Perspective. *The International Journal of Aviation Psychology.* 24(3): 190–209.

# 8 An Enhanced Approach in Basic Aircraft Maintenance Training Focusing on Human Factors

This chapter presents a proposal for the revision of the current EU Commission Regulation 1321/2014, especially Annex III (Part-66) and Annex IV (Part-147). The overall purpose of the suggested amendment to the regulation is to update the existing training system, in order to be capable to deliver skilled, well-educated and licensed aircraft maintenance engineers to the industry. It is envisaged that the adoption of the suggested approach will cover the declared needs of the aviation industry and respond to the challenges in the field of aircraft safety. This new approach attempts to integrate human factors into several modules both directly and horizontally, with the purpose of providing young engineers with the necessary skills and knowledge, thus enhancing and consolidating their attitude towards safety. The aim is to train them to recognize bad practices and norms that exist in maintenance organizations, whilst also enabling them to acknowledge and resist prevailing stressors that could affect their behavior, decisions and actions. It also aims to resolve the issue of the anticipated shortage of new engineers in the coming years, since it will attract more individuals to the profession and reduce the time needed for a candidate to become a licensed engineer without any negative impacts. Despite the fact that the proposal refers to the EASA's Annex III (Part-66) and Annex IV (Part-147) of EU Commission Regulation No. 1321/2014, it is envisaged that its main principles, scope and other training methodologies can be followed by other agencies.

## 8.1 CURRENT PATH TOWARDS THE AIRCRAFT MAINTENANCE LICENSE

In principle, the path to become an aircraft licensed engineer is similar among all agencies worldwide, since they all require a basic mandatory theoretical and practical training. However, there are deviations in the curricula, the methodology of examinations and especially as concerns the total duration of training that spans from three to four years up to five or six years, as depicted in Table 8.1.

Referring in particular to the European Union Member States and the third countries that have adopted EASA standards, aircraft maintenance engineers can issue a Certificate Release to Service (CRS) only if they have obtained their initial license

**TABLE 8.1**

**Worldwide Aircraft Maintenance License Requirements – An Indication**

| Area | Regulation | Basic Knowledge | Practical Experience | Authority Further Requirements | Average Time Needed (Consecutive – Full Time) | Title of Certificate |
|---|---|---|---|---|---|---|
| European Union | Commission Regulation (EU) No 1321/2014 Annex III (Part-66) | Demonstrate by examination a level of knowledge in the appropriate subject modules in accordance with Appendix I to Annex III (Part-66) | Depending on License Category (A, B1, B2, C) and previous experience and/or training followed. Spans between 1 to 5 years | Apply for license, no further exams | 4.5–5.5 years | Aircraft Maintenance License |
| US | 14 CFR Part 65 14 CFR Part 147 | 18 months of practical experience with either power plants or airframes, or 30 months of practical experience working on both at the same time or 2-year Aircraft Maintenance Technician Program approved by Federal Aviation Administration (FAA) | | FAA requires: written examination, oral test, practical test | 3–4 years | Aircraft Mechanic Certificate |
| Canada | Canadian Aviation Regulations, Part IV, Subpart 3 and Airworthiness Manual, Chapter 566 | Complete a Transport Canada-approved AME basic training program, usually 18-months duration | 4 years of experience | Transport Canada Civil Aviation requires technical and regulatory exam | 5–6 years | Aircraft Maintenance Engineer (AME) Licence |
| Asia (India) | Directorate General of Civil Aviation (DGCA) | 3–4 years engineering course and 1.5 year, 9 license exams conducted by Director General of Civil Aviation | 2 years of experience | Final license exam along with oral and practical test | 5–6 years | Aircraft Maintenance Engineer License |
| Australia | Part 66 of Civil Aviation Safety Regulations (CASR) | 4 years of apprenticeship to acquire practical experience and knowledge (technical and further education college, TAFE, 2 years of experience in category that license is for | | CASA requires two types of exams: core exams and specific group exams | 4.5–5.5 years | Licensed Aircraft Maintenance Engineer (LAME) |

*Source:* Created by Yiannakides, Sergiou 2019.

(basic license) from their competent authority, if they have completed a type rating on a given aircraft type endorsed in their license and if they have been authorized by the maintenance organization they work for.

Typically, engineers perform line and/or base maintenance and defect rectifications where:

- *Defect rectifications* refer to the activities that need to be undertaken by engineers to rectify a defect on an operational aircraft, in order to return it back to service. These can be simple tasks (e.g. replacement of electronic parts, simple skin repairs etc.), or complex tasks (e.g. autopilot or engine malfunctions) where the aircraft may be grounded for long periods, before returning back to service.
- *Line maintenance*, according to EASA Part 145, AMC 145.A.10, should be understood as "any maintenance that is carried out before flight to ensure that the aircraft is fit for the intended flight." In principle, typical line maintenance tasks would include a daily check (performed anywhere between 24 to 48 hours) and a weekly check (every 7–8 days). Apart from that, there may be several maintenance tasks which can be considered as line maintenance and carried out by a line maintenance provider. These may include:
  - Defect rectification
  - Simple troubleshooting
  - Replacement of line-replaceable unit (LRU), up to and including engines and propellers, with the use of external test equipment if required
  - Scheduled maintenance and/or checks, including visual inspections, that will detect obvious failures but do not require extensive in-depth inspection.
  - Inspection of systems and power plant items which are easily visible by opening access panels/doors
  - Minor repairs and modifications which do not require extensive disassembly and can be accomplished by simple means.

As analyzed in the previous chapters, aircraft engineers carry specific responsibilities that are associated with their license and the tasks involved, which have to be accomplished to the highest standards. Their working hours may become challenging since they are often required to work on weekends or night shifts. In order to effectively respond to the dynamic environment of aircraft maintenance, they should be self-reliant and self-disciplined, highly motivated, as well as capable of communicating with efficient housekeeping skills.

- *Base maintenance*, on the other hand, is all maintenance that does not fall under the line maintenance category. In practice, this mainly involves heavy checks such as C and D checks or even components failure. Nevertheless, the stressors related to the pressure that engineers experience when working in base maintenance are reduced, as compared to those experienced in line maintenance. Moreover, time pressure is also evident in base maintenance, mainly due to the global trend of reducing the time required to carry out maintenance in order to minimize cost and grounding periods. Taking into account that in base maintenance activities engineers have to deal with complex and time-consuming tasks on frequent occasions, such as corrosion prevention, structural work or replacement of major components, it is evident that other stressor factors such as mental workload and fatigue are also high. Likewise, in line maintenance, engineers have to work in night shifts and be competent and knowledgeable.

## 8.2 DRAWBACKS OF THE EXISTING SYSTEM (ANNEX III (PART-66) AND ANNEX IV (PART-147) OF COMMISSION REGULATION (EU) NO. 1321/2014)

As discussed in detail throughout this book, human factors are identified as a primary causal or directly contributing factor in aviation incidents/accidents. According to the *Evaluation report related to the EASA maintenance licensing system and maintenance training organizations* published in 2018, it is identified that:

- principles of SMS should be part of the basic knowledge syllabus (i.e. Part-66, Appendix I); and
- identification of hazards and the assessment of risks should be strictly observed during the imparting of the practical elements of training; notably:
  - introduce more HF, human performance and safety culture
  - attitude and corporate culture may explain fraud
  - focus on the areas of higher risks or areas of special emphasis.

The above considerations provide strong evidence that the existing training system needs improvement in various fields, such as human factors and human performance, as well as in other emerging sectors such as safety management system (SMS) and safety culture. They also consolidate the perception that unacceptable behaviors are evident in the aviation industry, since on several occasions experienced engineers deviate from best practices and follow bad norms that already exist within their organizations. Inevitably, such latent behaviors have the potential to become an infectious virus that taints newly employed and inexperienced engineers, who unconsciously adopt unsafe practices that they consider as correct, especially when these practices minimize the required time to perform a task and the mental and physical workload.

Based on the EASA observations above, the aviation community must strive to find solutions to achieve a better and more effective training for aircraft engineers. In this new approach, theoretical and *on the job training* on the impact of human factors on performance and safety should be a dominant factor and introduced directly and horizontally to basic knowledge (training modules).

As suggested in the beginning of this book, the need for a new training system is justified by the expected enlargement of the industry in the coming years and the anticipated enormous need for new competent and skilled engineers, who will be trained according to the highest standards. Also, the need for the new approach is bolstered by the fact that the industry expects an important labor shortage in the maintenance technician field. The reasons for the estimated shortage are several and range across different factors. According to ICAO these factors include the following:

- Wholesale retirements in the current generation of aviation professional
- Aviation professions not attractive enough to potential candidates
- Competition with other industry sectors for skilled employees
- Training capacity insufficient to meet demand
- Learning methodologies not responsive to new evolving learning style
- Accessibility to affordable training
- Lack of harmonization of competencies in some aviation disciplines, and little awareness by the "next generation" of types of aviation professions available.

When looking at the above factors presented by ICAO in detail, it is evident that the current training system itself is a significant factor contributing to the increase in the shortage of aviation engineers, either directly or indirectly. The majority of the reasons listed above are somehow related to training. Below, the most obvious are analyzed.

- *Aviation professions are not attractive enough to potential candidates:* It is a fact that there is a general lack of interest in aviation professions and especially in the field of engineering from younger talent pools. The aviation industry is no longer the attractive sector that it was during the previous decades. In a superficial approach, the causes can be attributed solely to the salaries and the hard-working conditions. However, this approach does not reflect the whole picture. Other reasons, such as the competition from alternative industries for skilled employees, high costs of education, affordability and accessibility are also dominant.
- *Competition with other industry sectors for skilled employees:* Skilled employees are valuable for all technical professions, especially in the aviation sector where tasks need to be performed in confined spaces, under high pressure and in most occasions within adverse working conditions. Unfortunately, the current training system is unable to produce skilled employees, whilst at the same time the system itself discourages skilled engineers from other technical professions to enter the aviation industry and thus acts as a disincentive.
- *Training capacity insufficient to meet demand:* The problem is intensified by the limited number of approved training centers, as well as the prolonged study period of an average of five years to receive a license.
- *Learning methodologies not responsive to new evolving learning style:* It is also a fact that the current methodologies are not responsive to the new evolving style. Moreover, the most worrisome factor is that the training material being taught in EASA Part-147 Aviation Training Centers requires

extensive reviews, in order to cope with the technological evolution of contemporary aircraft design and systems.

- *Accessibility to affordable training*: Affordable training is an issue in several types of professions. Besides the high cost, the initial training also suffers from the fact that students struggle to find maintenance organizations that are willing to offer them practical training upon completion of the theoretical training, while even when it is offered, it occurs under unpaid or very low-paid contracts.

## 8.3 PROBLEM STATEMENT

According to Annex III (Part-66) 66.A.25 (a) an applicant for an aircraft maintenance engineer license, or for the addition of a category or subcategory to such a license, shall demonstrate, by examination, a level of knowledge in the appropriate subject modules in accordance with Appendix I to Annex III (Part-66). The examination shall be conducted either by a training organization appropriately approved in accordance with Annex IV (Part-147) or by the competent authority.

Also, according to the same paragraph [Annex III (Part-66) 66.A.25 (a)], an applicant for an aircraft maintenance license for category B2 and subcategories B1.1 and B1.3 shall have acquired:

- 5 years of practical maintenance experience on operating aircraft if the applicant has no previous relevant technical training; or
- 3 years of practical maintenance experience on operating aircraft and completion of training considered relevant by the competent authority as a skilled worker, in a technical trade; or
- 2 years of practical maintenance experience on operating aircraft and completion of a basic training course approved in accordance with Annex IV (Part-147).

It is obvious based on the above, that candidates will need an average of five years of training to be able to apply for their Part-66 License.

Consequently, the questions raised are the following:

- Do the theoretical training and the exam system provide up-to-date knowledge to the candidates that render them capable to enter the industry as soon as possible?
- Is the academic background of these modular lessons adequate for this purpose?
- Does the practical training provide the necessary skills and competency to the candidates, so as to enter the industry as soon as possible?

As mentioned in the beginning of this book, surveys have already revealed the skills shortage in airline engineering. It is also a fact that criticism has emerged from the industry, stating that new engineers are not ready to work after almost five years of training. Consequently, they need to be trained again which results in massive resources being consumed from the companies.

### 8.3.1 POSSIBLE SCENARIO

To engage with the above questions, we present a possible scenario, within which applicants are hired by an EASA Part-145 certified organization as technicians (fitters), with no other education or academic background, apart from their high-school degree. Simultaneously, these individuals commence self-studying to pass the relevant examinations for the appropriate subject modules. According to the regulation, paragraph 66.A.25: "The training courses and examinations shall be passed within 10 years prior to the application for an aircraft maintenance license or the addition of a category or subcategory to such aircraft maintenance license." In such a scenario, it is normal for the priority of students to be the successful completion of the modules exams via any means. Although it is recognized that some knowledge could be gained, it cannot be compared with the knowledge acquired when students attend a typical or an interactive (online) class. Moreover, it can be argued that such tactics are favored, since the EASA module exams are structured in such a manner, with multiple-choice questions and just four short essays in total (two in Module 7 and one in Modules 9 and 10), the answers to which are often available on several websites. Also, in such cases, the students attend a number of "short courses" before the exams and memorize several questions in order to pass the exams.

Besides the obligation to pass the modular exams, applicants are required to provide proof of five or three years (skilled workers) of practical maintenance experience on operating aircraft. If the candidates already work in a Part-145 organization as technicians (fitters), they normally manage to receive all the necessary signatures to fill their logbook and submit it to the competent authority, in order to receive their initial license. Although this seems ideal, the reality is very much different. The engineers who work in Part-145 organizations are frequently under intense time pressure to deliver the aircraft and therefore the practical education of young engineers is not one of their first priorities. Thus, on frequent occasions, the practical experience becomes a "hunting" of signatures to fill the required tasks. It is also normal for students to be engaged in supplementary tasks like aircraft or hangar cleaning, or other simple tasks such as unscrewing panels, lubricating, greasing, counting tools, etc.

Evidently, the current training approach cannot "survive" in the context of future challenges, especially when the complexity of new aircraft and the dynamics of the maintenance environment are considered. The current options for receiving the B1 and/or B2 licenses are not adequate. Rather they simply deliver new engineers to the industry who lack skills and knowledge and are prone to errors. In addition, the required five years of practical experience do not constitute a structured education, since candidates are directly trained in real working conditions, where it is much more difficult to recognize bad norms, which they inevitably adopt as best practice.

The situation is definitely improved when students choose to attend a full course in an EASA Part-147 approved training school for 2–2.5 years and thereinafter an additional two years practical experience in an EASA Part-145 Maintenance Training Organization, to become eligible to apply for their EASA Part-66 license. This route guarantees that the candidates receive adequate theoretical training and that they can apply for the license with two more years of practical training.

However, the inefficiencies associated with the practical training, as explained above, remain the same.

In the next section an analysis is provided in an effort to identify the inefficiencies of the current training system and to suggest a new approach, aiming to respond to the industry's needs, in an effective and efficient manner. Concurrently, the suggested training system will reduce the average time needed for students to receive their B1/B2 license from the current five year period to at least three years. It is envisaged that this change will also reduce the overall training costs and act as a motivator to attract young, talented individuals to the profession of aircraft maintenance engineer.

## 8.4 PROPOSAL FOR THE REFORMATION OF EASA'S ANNEX III (PART-66) AND ANNEX IV (PART-147) OF COMMISSION REGULATION (EU) NO. 1321/2014

The introduction of an upgraded training and licensing model, based on Commission Regulation (EU) No. 1321/2014 of November 26, 2014 on the continuing airworthiness of aircraft and aeronautical products, parts and appliances, and on the approval of organizations and personnel involved in these tasks, is proposed, to meet the new aviation challenges. The proposed solution reforms Annex III (Part-66) and Annex IV (Part-147) of Regulation (EU) No. 1321/2014 (Basic Knowledge and Practical Experience).

### 8.4.1 BASIC KNOWLEDGE

Initially, a significant revision of the curriculum of the EASA Part-147 basic knowledge is proposed. The revision is needed in order to achieve efficiency through interaction and the participation of students in real life scenarios, as their knowledge progresses. Consequently, students will be offered the necessary up-to-date basic knowledge that will integrate human factors principles horizontally in all modules. It is also foreseen that the revision will educate students on the current trends and familiarize them with the advanced technologies of modern aircraft. To achieve the above, the following amendments are suggested:

- *Removal of Module 1 (Mathematics) and Module 2 (Physics).* These two modules along with the knowledge of the English language at a preselected level should be a prerequisite for anyone that wants to enter EASA's Part-147 training schools. Teaching basic mathematics (high-school level) and basic physics cannot be a task for an aviation training school. It should be a requirement for everyone who eventually wants to become a licensed aircraft engineer, as part of their basic knowledge, along with the required knowledge of the English language. The proper knowledge of basic mathematics and physics should be verified through dedicated exams (admission exam). For the verification of the English language level, it is suggested to establish a minimum global level with reference to the "Common European Framework of Reference for Languages."

- *Revision of the training material of Module 6 (Material and Hardware) and Module 7A (Maintenance Practices)* to meet the demands of modern aircraft and their technology improvements. For example, in Module 6 there are sub-modules that refer to wooden structures (6.3.2) and fabric covering (6.3.3), while in Module 7 the training material briefly covers computer-based techniques (uploading/downloading software to/from the aircraft).
- *Significant reduction in the training material of Modules 11A and 13.* It is suggested that these modules focus only on the basic principles concerning the operation of various aircraft systems (hydraulics, landing gears, central maintenance computers etc.) without specific details. Such an approach will reduce existing class hours and subsequently improve the knowledge on the operation/functions of fundamental aircraft systems.
- *Revision of Module 9A (Human Factors).* It is evident that according to the current EASA Training System, human factors training does not receive the attention it should. Human factors along with Module 8 (Basic Aerodynamics) are the shortest courses in this training. Frequently, the EASA Part-147 training organizations choose to teach human factors either first in the row or second after Module 8, since, due to their short length, they are considered ideal for the adaptation of new entrants in the Part-147 training school system. This tactic can to some extent be accepted, but it is evident that human factors, and their influence on maintenance activities, human attitude and decision-making, can be understood in a more efficient manner, both in class and through practical examples of the latent condition that may appear in this environment. Therefore, it is suggested that the first chapters of the existing curriculum (9.1 General, 9.2 Human Performance and Limitations, 9.3 Social Psychology) continue to be taught at the beginning of the program, while the remaining chapters are incorporated horizontally in the modules below:

  - Module 6 – Materials and Hardware
  - Module 7A – Maintenance Practices
  - Module 11A – Turbine Aeroplane Aerodynamics, Structures and Systems
  - Module 11B – Piston Aeroplane Aerodynamics, Structures and Systems
  - Module 12 – Helicopter Aerodynamics, Structures and Systems
  - Module 13 – Aircraft Aerodynamics, Structures and Systems
  - Module 15 – Gas Turbine Engine
  - Module 16 – Piston Engine
  - Module 17A – Propeller.

In each of the above modules, dedicated examples should be introduced, in order to alert students to the importance of following the procedures, reading the maintenance manuals and performing all the actions in an orderly and correct manner, without shortcuts. As an example, take Module 6, Submodule 6.5.4, that refers to Aircraft Rivets (*Types of Solid and Blind Rivets: Specification and Identification, Heat Treatment*) and Module 7, Submodule 7.9, that refers to Riveting (*Riveted Joints, Rivet Spacing Pitch, Tools Used for Riveting and Dimpling and Inspection and Clamping of Riveted Joints*). In the current system, there should normally be three–four multiple-choice questions in the exams for each submodule, requesting some specifications of the rivets, the proper distance between them or how to inspect them. Also, an essay could be introduced in Module 7, asking the students to explain the procedure of riveting on an aircraft in detail. However, the purpose of these courses should not only be to teach the typical procedures that are required to perform correct riveting, which is undoubtedly important for future engineers, but also to educate them on how to perform a very critical task like riveting on actual aircraft surfaces. Thus, the training should also focus on how to correctly follow the procedures and how to take appropriate safety measures according to the maintenance manuals and the instructions of the manufacturer. The importance of performing correct riveting will be imprinted forever in the heads of candidates, if the instructor refers to examples where bad riveting led to catastrophic events. A typical example that could be added as compulsory in the training material for this subsection could be the case study of Japan Airlines Flight 123 that crashed on August 12, 1985, resulting in a total of 520 deaths and four survivors. In that accident, virtually half of the tail fin had detached during flight, falling into the sea between Oshima Island and the headland. The investigation of the accident revealed that the same aircraft had been involved in a tail scrape during landing seven years earlier. The rear pressure bulkhead had been repaired at Osaka by Boeing Engineers. It was discovered that the pressure bulkhead had been incorrectly mended, with the doubler plate not extending across the entire repair, and as a result just a single line of rivets carried the loads. Thus, during the accident, the seam failed in the bulkhead, causing the bulkhead to blow out, creating an overpressure in the tail, severing the four sets of hydraulic control lines and blowing part of the tail section off.

The aforementioned case study could effectively integrate human factors principles in a horizontal manner to the module at hand and thus educate students to acknowledge the importance of following the procedures, the instructions and the maintenance manuals. Similarly, other case studies could be incorporated into the modules. Finally, exams in Module 9A (Human Factors) can be performed at the end of the training, with clearly enhanced content that will be justified by the knowledge acquired by the students.

- *Common Centralized Question Bank.* An important measure that will enhance the reliability and integrity of the exams and eliminate any fraud cases could be the development of a common centralized question bank. The question bank could be kept centrally by EASA and continually updated. The candidates will be examined only on questions from the training material and those questions will be selected randomly for each candidate in any exam. Paper-based exams should be allowed only for the essays.

### 8.4.2 PRACTICAL MAINTENANCE EXPERIENCE

The current requirements for practical maintenance experience are, as explained above, a significant obstacle to attracting both new students and technicians from other industries to the aviation industry. The problem, as already explained, is mostly due to the fact that students who have completed their basic training are required to join Part-145 maintenance organizations, in order to perform their practical training on operational aircraft and to sign their practical training logbooks, in order to be eligible to apply for the EASA Part-66 license. However, the reality is that Part-145 maintenance organizations are not training organizations and, as expected, they do not give the necessary attention, nor do they possess the required educational methods, individuals, experience or means.

Based on the above, it is proposed that the entire practical maintenance experience needed for the EASA Part-66 license be acquired within the EASA Part-147 organizations. This can be achieved by the implementation of a "mock (mirror) training Part-145 organization," which will be run by authorized instructors. Students will be eligible to enter the practical training after they complete all the theoretical modules, or part of them. The aim is to offer students standardized practical experience. This approach in training will integrate safety culture and in particular familiarize students with the basic principles of human factors, as well as provide them with enhanced maintenance skills. Also, this approach will reduce the total time required for obtaining the license, as it will be provided in a coherent and uninterrupted manner. In order to implement this approach, a basic requirement could be for each EASA Part-147 training organization performing a full basic course to maintain a small functional aircraft of any type (e.g. CESSNA 150, 172, Diamond etc.), upon which the students could perform all the required tasks.

The fact that through this solution, the engineers may receive their Part-66 license without interaction with the real maintenance environment (Part-145), is counterbalanced by the fact that according to 66.A.45(c), "the endorsement of the first aircraft type rating within a given category/sub-category requires satisfactory completion of the corresponding On the Job Training (OJT), as described in Appendix III to Annex III (Part-66)." This allows the new engineers enough time, after they receive their initial Part-66 license and before the endorsement of the first type, to interact with the real maintenance environment (Part-145) during their individualized OJT training, but under an enhanced practical knowledge base in comparison with the existing situation. The practical experience received though the proposed system will allow the engineers to quickly adapt and finish their OJT training, in much less time and more effectively, while it will render them capable to recognize bad norms and follow the indicated procedures.

In particular, through the proposed system and further to the practical experience gained during the theoretical modules, the students will acquire the following skills and knowledge:

- Proper use of basic tools like torque wrenches, multimeters, bonding testers, compressed air tools, drilling, riveting tools, etc. on the aircraft, using the technical publication of the manufacturer and the standard practice manuals.
- Proper and efficient use of these tools is one of the basic skills that a new engineer should possess, but which require substantial practice. Practical experience within the Part 147 organizations under the new approach will be tolerant to errors, delays and system damage, and also free of

any economic repercussions that emerge when practical training is conducted in real maintenance environments. Instructors can focus on the absolute safety measures that should be taken, both for the personnel and the equipment, which is a condition often overlooked in real maintenance environments.

- Dedicated training in novel skills like software updates, computer base fault finding, etc.
- Understanding and use of maintenance manuals of all kinds. This is one of the major problems that new engineers face when they join Part-145 organizations. In the proposed system, the practical use of maintenance manuals on a functional demo airplane, as part of their training, can render the students capable to fully understand the procedure without time pressure. Also, fault solving procedures can enhance this process.
- Understanding and implementing the airworthiness directives, service bulletins, service letters, etc. The students can go through all the procedures and perform these tasks on the airplane.
- Practical understanding of the legislation, for example, training on storage procedures such as the receiving of functional parts and the delivery of unserviceable parts to storage, issuing of Certificate Release to Service (CRS), etc.
- Practical understanding of human factors. Further to the teaching of Module 9A in the initial course, there is also a mandatory requirement for recurrent training for all maintenance personnel as outlined in EASA 145.A.30(e). In addition, the *Evaluation report related to the EASA maintenance licensing system and maintenance training organizations*, published on March 2, 2018, identifies that other elements should also be taken on board, such as the principles of safety management system (SMS) as well as hazard identification and risk assessment, human performance and safety culture.

Based on the above findings, it is suggested that part of the practical training should be devoted to simulating scenarios that may occur in real maintenance environments and which have the potential to lead to unsafe acts. Typical scenarios may include the impact on performance and the consequent safety actions when engineers are disarrayed or interrupted by external factors, such as mobile phone ringing, loud noise, colleagues' conversation etc., or overloaded by multiple tasking and time pressure. Also, scenarios may include other significant elements such as error identification, lack of communication, team decision-making, shifting team briefings etc.

- Final assessment of students' practical experience and skills may be performed by an external assessor either from the EASA Part-147 organization or even better by the competent authority.

## 8.5 STANDARDIZATION

Further standardization between EASA and Part-147 training organizations will ensure that the training standards are followed independently. To this end, the ideas presented below are not part of the suggested proposal, however they can be considered as "food for thought" for future updates of the training system.

- *Training related to effective jobs/task* on the aircraft, such as changing a propeller, weight and balance and Non-Destructive Testing (NDT) to be undertaken in the following two ways: a part using special simulators (simulating real actions as implemented for pilots) and the other part (which cannot be completed with simulators) to be undertaken on a functional aircraft, in partnership with an approved Part-145 Maintenance Organization, *but under the supervision and responsibility of the Part-147 organization.*
- *Common/central curriculum (training material):* A step that leads to standardization is the development of a center/organization that will be responsible to create and deliver the same basic training material to all EASA Part-147 Training Schools (same books, electronic material etc.). Currently, each EASA Part-147 training school delivers its own training material to the candidates, typically approved by the competent authority. The training material varies between different EASA Part-147 training centers, since their only common denominator is the titles of the modules and submodules that are referred to in the regulation and their levels of training. Based solely on this, each training center develops its own training material. The development of an overarching *center/organization* responsible to deliver common basic training material to all EASA Part-147 Training Centers will ensure that these institutions teach at least this common training material. This center/organization will centrally receive feedback from all the EASA Part-147 Training Centers around the globe and amend the training material accordingly. It will be responsible to monitor the evolution of the industry and adapt the training material centrally. Thus, every training center will always be up-to-date and in full alignment with the remaining centers. Also, a measure like this, will allow EASA to easily update the training materials in a global manner and adapt rapidly to any new requirements emanating from technological or other improvements. It is obvious that measures of this kind will immediately elevate the training standards of EASA Part-147 organizations.
- *Correlation with academic institutions:* Recalling ICAO factors concerning the shortage of skilled aviation professionals, the following factors (briefly analyzed above) can be further examined from the training point of view:
  - *Aviation professions not attractive enough to potential candidates*
  - *Competition with other industry sectors for skilled employees.*

As regards the second point, competition with other industry sectors for skilled employees, it can be easily concluded that the current training system and the existing requirements in the regulation are not attractive enough for engineers from other sectors to join the aviation sector. Since the basic theoretical knowledge is required for every student, in practice this implies that an engineer from any other industry sector needs to pass all the modular exams, or at least the vast majority of them (depending on the Competent Authority), whilst also demonstrating three years of practical maintenance experience on operating aircraft, in order to prove that he/she is a "skilled worker" according to the regulation. *A "skilled worker" is a person who has successfully completed training acceptable to the competent authority and involving the manufacture, repair, overhaul or inspection of mechanical, electrical or electronic equipment. The training would include the use of tools and measuring devices.*

Based on this observation, it becomes obvious that it is very difficult for professional engineers currently employed in other industries to switch to the aviation industry and especially to the aircraft maintenance sector. This is an important reason, among others, that explains the observation by ICAO that the "aviation professions [are] not attractive enough to potential candidates."

Concurrently, the same training system makes it very hard for young, talented high-school students to continue their studies in Part-147 training schools. This is mainly due to the fact that the product of the studies is the EASA Part-66 License, which, as it stands now, has no correspondence with college or university degrees, nor is it accredited as an academic certificate. Subsequently, when students who have already been enrolled in the basic course decide to abandon it, all accomplishments up to that point have no value, nor do they provide any established prospective that will facilitate students to join other colleges or universities in the engineering fields.

To mitigate the effects of this issue and render the aviation industry more attractive, it is proposed for the enhanced training system be incorporated into the European Credit Transfer and Accumulation System (ECTS). This will allow students to continue their studies and receive graduate or postgraduate degrees, whilst also attracting students from other universities to integrate Part-66 training under advantageous conditions such as the exception from modules, less funds etc. Nevertheless, the common standards between all EASA Part-147 organizations, as previously analyzed, are fundamental to this effort.

# Index

Page numbers in **bold** denote tables.